Nanomaterial-Based Emerging Technologies for Detecting Food Contaminants

Nanomaterial-Based Emerging Technologies for Detecting Food Contaminants

Editors

Mingfei Pan
Longhua Xu
Huilin Liu

MDPI

Basel • Beijing • Wuhan • Barcelona • Belgrade • Novi Sad • Cluj • Manchester

Editors
Mingfei Pan
Tianjin University of Science and Technology
Tianjin
China

Longhua Xu
Shandong Agricultural University
Taian
China

Huilin Liu
Beijing Technology and Business University
Beijing
China

Editorial Office
MDPI
St. Alban-Anlage 66
4052 Basel, Switzerland

This is a reprint of articles from the Special Issue published online in the open access journal *Foods* (ISSN 2304-8158) (available at: https://www.mdpi.com/journal/foods/special_issues/nanomaterial_food_contaminants).

For citation purposes, cite each article independently as indicated on the article page online and as indicated below:

Lastname, A.A.; Lastname, B.B. Article Title. *Journal Name* **Year**, *Volume Number*, Page Range.

ISBN 978-3-7258-0595-2 (Hbk)
ISBN 978-3-7258-0596-9 (PDF)
doi.org/10.3390/books978-3-7258-0596-9

Contents

foods

Editorial

Nanomaterial-Based Optical Detection of Food Contaminants

Mingfei Pan

State Key Laboratory of Food Nutrition and Safety, Tianjin University of Science and Technology, Tianjin 300457, China; panmf2012@tust.edu.cn

1. Introduction

The presence of food contaminants remains a significant aspect contributing to global food safety issues, drawing widespread attention from ordinary consumers, governments, and researchers [1–3]. These contaminants encompass various harmful factors, including residues of agricultural and veterinary drugs, biotoxins, heavy metals, allergenic proteins, and particularly endogenous hazardous substances generated during processing, representing a new focus in detection technology research [4,5].

Various nanomaterials with different structures and properties, such as metallic nanomaterials, up-conversion fluorescence nanomaterials, metal–organic framework porous materials, quantum dot fluorescence materials, etc., not only serve as solid carriers for biorecognition elements (such as antibodies, aptamers, etc.) [6,7] and biomimetic recognition elements (such as molecularly imprinted polymers) [8,9], but also provide signal sources for visual, rapid, and convenient analysis [10–12]. Visible light, fluorescence, or electrochemiluminescence are quantitative signals commonly used in convenient food safety analysis strategies. This significantly promotes the development of precise and rapid analysis techniques for food contaminants, incorporating advanced methods such as nanomaterials, biomimetics and biorecognition, and chemometrics. Therefore, strategies based on multifunctional nanomaterials and utilizing antibodies, aptamers, and biomimetic polymers as recognition elements for the fluorescent or visual detection of contaminants in complex food matrices are gradually assuming crucial roles in food safety testing strategies [13–15].

2. An Overview of Published Articles

In the study by Tianyu Ma et al. (Contribution 1), the authors designed and prepared a zearalenone (ZEN) hapten against the mycotoxin ZEN, and the original coating ZEN-ovalbumin (ZEN-OVA) by conjugation with OVA. Based on gold nanorods (AuNRs) of uniform size and stable properties synthesized by the seed-mediated method, the indirect competitive enzyme-linked immunosorbent assay (ic-ELISA) and the AuNR growth-based multicolor ELISA for detecting ZEN toxin were further established. Under optimal experimental conditions, the coating amounts of ZEN-OVA were 0.025 μg/well, the antibody (Ab) dilution factor was 32,000 times, blocking solution was 0.5% skimmed milk powder, enzyme-labeled secondary Ab diluted 10,000 times, and at pH 7.4 of the PBS buffer, the sensitivity (IC_{50}) of the established ic-ELISA for ZEN detection reached 0.85 ± 0.04 μg/L and the limit of detection (LOD, IC_{15}) reached 0.22 ± 0.08 μg/L. In the multicolor ELISA based on the growth of AuNRs, as the content of ZEN increased, the mixed solution exhibited a significant color change from brownish red to colorless. ZEN concentrations as low as 0.1 μg/L could be detected with the naked eye (brown-red to dark gray). This study presents an effective analysis strategy for the rapid screening and accurate monitoring of ZEN contaminants in foods.

In the study by Lingyan Zhao et al. (Contribution 2), a novel rare earth upconversion nanomaterial with a three-layer sandwich core–shell structure was synthesized by an improved thermal decomposition method, and the morphology, fluorescence intensity, and diffraction peak position of the new material were characterized by TEM (transmission

Citation: Pan, M. Nanomaterial-Based Optical Detection of Food Contaminants. Foods 2024, 13, 557. https://doi.org/10.3390/foods13040557

Received: 26 January 2024
Accepted: 3 February 2024
Published: 12 February 2024

electron microscopy), XRD (powder X-ray diffraction), and fluorescence spectrophotometry. The inert core/active shell/inert shell design improved the upconversion luminous efficiency of the new material several-fold. FT-IR (Fourier transform infrared spectroscopy) characterization showed that the surface of activated upconversion nanoparticles was modified with the silicon shell and amino group. Combined with the characteristics that aminoated polystyrene magnetic microspheres could be separated by the magnetic field, an upconversion magnetic separation immunoassay method for the detection of pyrethroid pesticide residues was established. The capture probe competed with the pyrethroid standard, combined the signal probe, and measured the fluorescence signal value formed by the capture probe signal probe complex at 542 nm under 980 nm excitation light. The LOD of fenpropathrin was 0.01 μg/L, cypermethrin was 0.015 μg/L, and fenvalerate was 0.011 μg/L. Through the actual detection of apple, cabbage, and other samples, the recovery rate of pyrethroids was between approximately 83.4% and 97.8%. Comparison with the HPLC (high-performance liquid chromatography) detection results showed that the established method had good accuracy and could realize the quantitative analysis of pyrethroids in food.

In their research, Wenbo Zhu et al. (Contribution 3) coupled two upconversion materials with anti-clothianidin and anti-imidacloprid monoclonal antibodies as signal probes using the glutaraldehyde cross-linking method. Under the excitation of 980 nm excitation light, the fluorescence signals of the synthesized core–shell $NaYF_4$:Yb@$NaYF_4$:Ho and monolayer $NaYF_4$:Yb,Tm upconversion nanoparticles (UCNPs) were simultaneously detected at 656 and 696 nm, respectively. Imidacloprid (IMI) and clothianidin (CLO) could compete with antigen-conjugated amino Fe_3O_4 magnetic nanomaterials for binding to signaling probes, thus establishing a rapid and sensitive fluorescent immunoassay for the simultaneous detection of IMI and CLO. Under optimal conditions, the LOD (IC_{10}) and sensitivity (IC_{50}) of IMI and CLO were (0.032, 0.028) and (4.7, 2.1) ng/mL, respectively, and the linear assay ranges were at 0.032–285.75 ng/mL and 0.028–200 ng/mL, respectively. The immunoassay did not significantly cross-react with other analogs. In fruits and vegetables such as apples, oranges, peaches, cucumbers, tomatoes, and peppers, the mean recoveries of IMI and CLO ranged from 83.33% to 115.02% with relative standard deviations (RSDs) of 1.9% to 9.2% and 1.2% to 9.0%, respectively. Furthermore, the results of the immunoassay correlate well with the high-performance liquid chromatography method used to detect the actual samples.

In the work of Chang Liu et al. (Contribution 4), a Eu^{3+}-MOF-253@Au electrochemiluminescence sensor was successfully constructed for the first time by encapsulating nanogold in the metal–organic framework (MOF) backbone and pore channels, and assembling Eu^{3+} on the MOF backbone. Firstly, the introduction of nanogold overcame the weakness of MOFs, which was difficult to achieve, and enhanced its catalytic performance, followed by the modification of Eu^{3+} to confer the electrochemiluminescence performance and the function of target detection on the sensor. Moreover, carbaryl was placed in an alkaline working solution to enhance the intensity of electrochemiluminescence signals, as well as to promote the hydrolysis of carbaryl into 1-naphthol, which caused the burst of the Eu^{3+}-MOF-253@Au electrochemiluminescence sensor, thereby achieving the sensitive detection of carbaryl. On this basis, the electrochemiluminescence detection conditions were optimized, the performance was analyzed, and finally, it was successfully used for the detection of carbaryl with good linearity in the range of 0.2–200 μg L^{-1} and a low LOD (0.14 μg L^{-1}).

Food allergies have seriously affected some people's quality of life, and even endangered their lives. At present, there is still no effective cure for food allergies. Avoiding the intake of allergenic food is still the most effective way to prevent allergic diseases. Therefore, it is necessary to develop rapid, accurate, sensitive, and reliable analysis methods to detect food allergens from different sources. Aptamers are oligonucleotide sequences that can bind to a variety of targets with high specificity and selectivity, and they are often combined with different transduction technologies, thereby constructing various types of aptamer sensors. In recent years, with the development of technology and the application of new materials, the sensitivity, portability, and cost of fluorescence sensing technology have been greatly improved. Therefore, aptamer-based fluorescence sensing technology has been widely developed and applied in the specific recognition of food allergens. Liping Hong et al. (Contribution 5) comprehensively reviewed the classification of major allergens and their characteristics in animal and plant foods, and summarized the preparation principles and practical applications of aptamer-based fluorescence biosensors. This article presents some strategies for the rapid and sensitive detection of allergens in food matrices.

In their study, Ziwen Zhang et al. (Contribution 6) developed a rapid fluorescent and colorimetric dual-mode detection strategy for Hg^{2+} in seafoods based on the cyclic binding of the organic fluorescent dye rhodamine 6G hydrazide (R6GH) to Hg^{2+}. The luminescence properties of the fluorescent R6GH probe in different systems were investigated in detail. Based on the UV and fluorescence spectra, it was determined that the R6GH has good fluorescence intensity in acetonitrile and good selective recognition of Hg^{2+}. Under optimal conditions, the R6GH fluorescent probe showed a good linear response to Hg^{2+} ($R^2 = 0.9888$) in the range of 0–5 μM, with a low detection limit of 2.5×10^{-2} μM (S/N = 3). A paper-based sensing strategy based on fluorescence and colorimetric analysis was developed for the visualization and semiquantitative analysis of Hg^{2+} in seafoods. The LAB values of the paper-based sensor impregnated with the R6GH probe solution showed good linearity ($R^2 = 0.9875$), with Hg^{2+} concentrations in the range of 0–50 μM, which means that the sensing paper could be combined with smart devices to provide reliable and efficient Hg^{2+} detection.

Additives and antibiotic abuse during food production and processing are among the key factors affecting food safety. The efficient and rapid detection of hazardous substances in food is of crucial relevance to ensure food safety. In the study by Shijie Li et al. (Contribution 7), a water-soluble quantum dot with glutathione as a ligand was synthesized as a fluorescent probe by the hydrothermal method to achieve the detection and analysis of H_2O_2. The detection limits were 0.61 μM in water and 68 μM in milk. Moreover, it was used as a fluorescent donor probe, and manganese dioxide nanosheets were used as a fluorescent acceptor probe in combination with an immunoassay platform to achieve the rapid detection and analysis of enrofloxacin (ENR) in a variety of foods with LODs of 0.05–0.25 ng/mL in foods. The proposed systems provide new ideas for the construction of fluorescence sensors with high sensitivity.

The study from Qi Zhang et al. (Contribution 8) proposes a facile and versatile layer-by-layer strategy without any special surface modifications for the preparation of magnetic metal-organic frameworks (MMOFs) supporting molecularly imprinted polymer nanoparticles (MMOFs@MIP), which are based on a magnetically susceptible core conjugated with an imidazole-derived self-assembled layer and a silane-based imprinted shell. Metal–organic frameworks (MOFs) with systematically tailored structures have been suggested as promising precursors to the preparation of diverse functional materials. The obtained MMOFs@MIPs, which integrated the advantages of Fe_3O_4, MOFs, and MIPs, were characterized and exhibited good magnetic properties, a rapid mass transfer rate, and excellent adsorption selectivity, as well as capacity for the targeted molecular—bisphenol A (BPA). Moreover, the MMOFs@MIPs were employed as adsorbents in magnetic solid-phase extraction (MSPE) to selectively bind and rapidly separate BPA from real samples, with satisfactory recovery rates ranging from 88.3% to 92.3%. More importantly, the desirable reusability of MMOFs@MIPs was also evaluated, and the recovery was maintained above 88.0% even after five re-use cycles. Furthermore, combined with high-performance liquid chromatography (HPLC) analysis, a novel MSPE-HPLC method was developed, enabling the highly selective and sensitive detection of BPA in a wide linear range of 0.5–5000 μg L^{-1}, with a low LOD of 0.1 μg L^{-1}. This work contributes a promising method for constructing various functional nanoparticle@MOFs@MIP hybrid materials for applications in many different fields.

Ying Guo et al. constructed a novel fluorescent molecularly imprinted nanosensor (N, S-GQDs@ZIF-8@MIP) based on the nitrogen and sulfur co-doped graphene quantum dots decorated zeolitic imidazolate framework-8 for the detection of octopamine (OA) (Contribution 9). Herein, ZIF-8 with a large surface area was introduced as a supporter of the sensing system, which effectively shortened the response time of the sensor. Meanwhile, high green luminescent N, S-GQDs and a maximum emission wavelength of 520 nm under 460 nm excitation and a 12.5% quantum yield, were modified on the surface of ZIF-8 as a signal tag that could convert the interactions between the sensor and OA into detectable fluorescent signals. Finally, N, S-GQDs@ZIF-8@MIP was acquired through the surface molecular imprinting method. Due to the synergy of N, S-GQDs, ZIF-8, and MIP, the obtained sensor not only demonstrated higher selectivity and sensitivity than N, S-GQDs@ZIF-8@NIP, but also displayed faster fluorescence responses than N, S-GQDs@MIP. Under optimal conditions, the developed sensor presented a favorable linear relationship in the range of 0.1–10 mg L^{-1}, with a detection limit of 0.062 mg L^{-1}. Additionally, the proposed N, S-GQDs@ZIF-8@MIP strategy was effectively applied to the detection of OA in fermented samples.

Xiaohui Wang et al. constructed a fluorescent sensor (NH_2-UIO-66(Zr)@MIP) based on a molecularly imprinted polymer-coated amino-functionalized zirconium (IV) metal–organic framework, and initially used it for the ultrasensitive determination of oxytetracycline (Contribution 10). Developing sensitive and effective methods to monitor oxytetracycline residues in food is of great significance for maintaining public health. NH_2-UIO-66 (Zr), with a maximum emission wavelength of 455 nm under 350 nm excitation, was prepared using a microwave-assisted heating method. The NH_2-UIO-66(Zr)@MIP sensor with specific recognition sites for oxytetracycline was then acquired by modifying a molecularly imprinted polymer on the surface of NH_2-UIO-66 (Zr). The introduction of NH_2-UIO-66 (Zr) as both a signal tag and supporter could strengthen the sensitivity of the fluorescence sensor. Thanks to the combination of the unique characteristics of the molecularly imprinted polymer and NH_2-UIO-66 (Zr), the prepared sensor not only exhibited a sensitive fluorescence response, specific identification capabilities, and a high selectivity for oxytetracycline, but also showed good fluorescence stability, satisfactory precision, and reproducibility. The fabricated sensor displayed fluorescent linear quenching in the OTC concentration range of 0.05–40 μg mL^{-1}, with a detection limit of 0.012 μg mL^{-1}. More importantly, the fluorescence sensor was finally applied for the detection of oxytetracycline in milk, and the results were comparable with those obtained using the HPLC approach. Hence, the NH_2-UIO-66 (Zr)@MIP sensor possesses great application potential for the accurate evaluation of trace oxytetracycline in dairy products.

Currently, detection technologies regarding food contaminants are still facing multiple challenges, which include issues such as the effective removal of food matrices and the monitoring and control of hazardous substances generated during food processing. It is believed that the development of more efficient, accurate, and sensitive strategies for the detection and control of food contaminants will be gradually promoted with the continuous advancement of nanomaterials, food-related research, and chemometrics. The aim of this Special Issue is to publish high-quality articles on the accurate and rapid detection of various types of food contaminants in the areas of fluorescent nanomaterials, biometrics and biomimetic recognition, and fluorescent molecules based on organic dyes and quantum dots, in order to promote the further development of nanomaterial-based optical sensing and detection methods.

Conflicts of Interest: The author declares no conflicts of interest.

References

1. Fung, F.; Wang, H.S.; Menon, S. Food safety in the 21st century. *Biomed. J.* **2018**, *41*, 88–95. [CrossRef] [PubMed]
2. Gallo, M.; Ferrara, L.; Calogero, A.; Montesano, D.; Naviglio, D. Relationships between food and diseases: What to know to ensure food safety. *Food Res. Int.* **2020**, *137*, 109414. [CrossRef] [PubMed]

3. Shen, Y.; Wei, Y.; Zhu, C.; Cao, J.; Han, D.M. Ratiometric fluorescent signals-driven smartphone-based portable sensors for onsite visual detection of food contaminants. *Coordin. Chem. Rev.* **2022**, *458*, 214442. [CrossRef]
4. World Health Organization. *Safety Evaluation of Certain Contaminants in Food: Prepared by the Ninetieth Meeting of the Joint FAO/WHO Expert Committee on Food Additives (JECFA)*; WHO: Geneva, Switzerland, 2023.
5. Chen, J.; Sun, R.; Pan, C.; Sun, Y.; Mai, B.; Li, Q.X. Antibiotics and food safety in aquaculture. *J. Agric. Food Chem.* **2020**, *68*, 11908–11919. [CrossRef] [PubMed]
6. Ma, T.; Liu, K.; Yang, X.; Yang, J.; Pan, M.; Wang, S. Development of indirect competitive ELISA and visualized multicolor ELISA based on gold nanorods growth for the determination of zearalenone. *Foods* **2021**, *10*, 2654. [CrossRef] [PubMed]
7. Hong, L.; Pan, M.; Xie, X.; Liu, K.; Yang, J.; Wang, S.; Wang, S. Aptamer-based fluorescent biosensor for the rapid and sensitive detection of allergens in food matrices. *Foods* **2021**, *10*, 2598. [CrossRef] [PubMed]
8. Villa, C.C.; Sánchez, L.T.; Valencia, G.A.; Ahmed, S.; Gutiérrez, T.J. Molecularly imprinted polymers for food applications: A review. *Trends Food Sci. Technol.* **2021**, *111*, 642–669. [CrossRef]
9. Ayerdurai, V.; Cieplak, M.; Kutner, W. Molecularly imprinted polymer-based electrochemical sensors for food contaminants determination. *TrAC-Trend. Anal. Chem.* **2023**, *158*, 116830. [CrossRef]
10. Hua, Z.; Yu, T.; Liu, D.; Xianyu, Y. Recent advances in gold nanoparticles-based biosensors for food safety detection. *Biosens. Bioelectron.* **2021**, *179*, 113076. [CrossRef] [PubMed]
11. Umapathi, R.; Park, B.; Sonwal, S.; Rani, G.M.; Cho, Y.; Huh, Y.S. Advances in optical-sensing strategies for the on-site detection of pesticides in agricultural foods. *Trends Food Sci. Technol.* **2022**, *119*, 69–89. [CrossRef]
12. Zhang, Z.; Lou, Y.; Guo, C.; Jia, Q.; Song, Y.; Tian, J.Y.; Zhang, S.; Wang, M.; He, L.; Du, M. Metal–organic frameworks (MOFs) based chemosensors/biosensors for analysis of food contaminants. *Trends Food Sci. Technol.* **2021**, *118*, 569–588. [CrossRef]
13. Umapathi, R.; Sonwal, S.; Lee, M.J.; Rani, G.M.; Lee, E.S.; Jeon, T.J.; Kang, S.M.; Oh, M.H.; Huh, Y.S. Colorimetric based on-site sensing strategies for the rapid detection of pesticides in agricultural foods: New horizons, perspectives, and challenges. *Coordin. Chem. Rev.* **2021**, *446*, 214061. [CrossRef]
14. Nile, S.H.; Baskar, V.; Selvaraj, D.; Nile, A.; Xiao, J.; Kai, G. Nanotechnologies in food science: Applications, recent trends, and future perspectives. *Nano-Micro Lett.* **2020**, *12*, 45. [CrossRef] [PubMed]
15. Barzegar, F.; Kamankesh, M.; Mohammadi, A. Recent development in formation, toxic effects, human health and analytical techniques of food contaminants. *Food Rev. Int.* **2023**, *39*, 1157–1183. [CrossRef]

foods

Article

Development of Indirect Competitive ELISA and Visualized Multicolor ELISA Based on Gold Nanorods Growth for the Determination of Zearalenone

Tianyu Ma [1,2], Kaixin Liu [1,2], Xiao Yang [1,2], Jingying Yang [1,2], Mingfei Pan [1,2,*] and Shuo Wang [1,2]

1 State Key Laboratory of Food Nutrition and Safety, Tianjin University of Science & Technology, Tianjin 300457, China; maty1128@126.com (T.M.); Liukx2019@163.com (K.L.); yangx2021@126.com (X.Y.); yangjy0823@126.com (J.Y.); s.wang@tust.edu.cn (S.W.)
2 Key Laboratory of Food Nutrition and Safety, Ministry of Education of China, Tianjin University of Science and Technology, Tianjin 300457, China
* Correspondence: panmf2012@tust.edu.cn; Tel.: +86-022-6091-2493

Abstract: In this study, a zearalenone (ZEN) hapten was designed and prepared against the mycotoxin ZEN, and the original coating ZEN-ovalbumin (ZEN-OVA) was prepared by conjugation with OVA. Based on the gold nanorods (AuNRs) of uniform size and stable properties synthesized by the seed-mediated method, the indirect competitive enzyme-linked immunosorbent assay (ic-ELISA) and the AuNRs growth-based multicolor ELISA for detecting ZEN toxin were further established. Under the optimal experimental conditions, the coating amount of ZEN-OVA: 0.025 μg/well, antibody (Ab) dilution factor: 32,000 times, blocking solution: 0.5% skimmed milk powder, enzyme-labeled secondary Ab diluted 10,000 times, and a pH of the PBS buffer at 7.4, the sensitivity (IC_{50}) of the established ic-ELISA for ZEN detection reached 0.85 ± 0.04 μg/L, and the limit of detection (IC_{15}) reached 0.22 ± 0.08 μg/L. In the multicolor ELISA based on the growth of AuNRs, as the content of ZEN increased, the mixed solution exhibited a significant color change from brownish red to colorless. ZEN concentration as low as 0.1 μg/L can be detected by the naked eye (brown red to dark gray). This study provided an effective analysis strategy for the rapid screening and accurate monitoring of the ZEN contaminant in foods.

Keywords: zearalenone; gold nanorods; indirect competitive ELISA; visualized multicolor ELISA

Citation: Ma, T.; Liu, K.; Yang, X.; Yang, J.; Pan, M.; Wang, S. Development of Indirect Competitive ELISA and Visualized Multicolor ELISA Based on Gold Nanorods Growth for the Determination of Zearalenone. *Foods* **2021**, *10*, 2654. https://doi.org/10.3390/foods10112654

Academic Editor: Andrew G. Gehring

Received: 6 October 2021
Accepted: 28 October 2021
Published: 2 November 2021

1. Introduction

Zearalenone (ZEN) is a nonsterol estrogen mycotoxin mainly produced by *Fusarium* genera, which is widely found in grain crops, such as corn, wheat, and barley [1,2]. The degree of contamination of food crops by *Fusarium* mainly depends on the moisture and temperature conditions; the optimal growth temperature for *Fusarium* is between 24 and 32 °C and the optimal humidity is 40% [3]. Therefore, in areas with sufficient rainfall and high relative humidity, cereals are likely to be contaminated with *Fusarium* in all steps of production, storage, and processing [4]. ZEN can cause excessive estrogen syndrome in animals such as pig, poultry, and humans, as well as immunotoxicity, genotoxicity, and suspected carcinogenicity [5,6]. It has been reported that the potential etiological mechanism of breast cancer involves changes in the cytochrome P450 (CYP) enzyme, which is related to ZEN [7]. Yu et al. have demonstrated that the ZEN on *2,3,7,8*-Tetrachlorodibenzo-*p*-dioxin (TCDD)-induced CYP1A1 activity and gene expression involved the estrogen receptor pathway [8]. In addition, due to its similar structure to endogenous estrogen, ZEN can show estrogen activity in vivo and competitively bind with estrogen receptors, thus affecting estrogen secretion in humans or animals, resulting in reproductive organ abnormalities, infertility, abortion, and other diseases [9–11]. At present, many countries and organizations have regulated the maximum residue levels (MRLs) of ZEN in foods, although a consistent standard has not yet been obtained (European Union: 20–400 μg/kg in different types of foods, Russia: 1000 μg/kg in hard wheat, flour and wheat germ; China: 60 μg/kg in cereals and its products) [12–14]. Therefore, the development of effective strategies for ZEN detection in foods is of great significance for protecting the health of humans and animals.

Currently, traditional instrumental methods based on liquid chromatography (LC), gas chromatography (GC), and mass spectrometry (MS) have been widely applied in the detection of ZEN contaminant in foods and animal feed samples [15–20]. However, these techniques require sophisticated instruments, professional and experienced operators, and long test time, which limits the practical application to a certain extent. Engvall and Perlmann achieved the quantitative detection of a solid phase enzyme immunoassay for the first time, marking the successful construction of an enzyme linked immunosorbent assay (ELISA) [21,22]. The ELISA has the characteristics of simple operation, high specificity, and low cost, and has gradually become the mainstream technology in the field of rapid detection [23,24]. Improving the sensitivity and lowering the detection limit of ELISA can promote the further application of this effective technology [25,26]. Signal amplification has been demonstrated by many researchers to improve the sensitivity of ELISA. In particular, the introduction of various nanomaterials has further improved the performance of traditional ELISA. For example, Zhang et al. synthesized CdTe/CdS/ZnS quantum dots (QDs) in the aqueous phase and developed a fluorescent immunoassay (FLISA) for detecting ZEN in corn [27]. Xiong et al. reported an advanced enzyme-assisted etching method in which gold nanorods (AuNRs) were applied as the signal carrier for aflatoxin B_1 (AFB_1) in corn samples to amplify the ELISA signal [28]. Liu et al. constructed a horseradish peroxidase (HRP)-mediated ratio fluorescence ELISA based on gold and silver bimetallic nanoclusters (Au-AgNCs) to detect zearalenone, which significantly improved the detection limit [29].

In this study, we successfully prepared the ZEN hapten and ZEN coating antigen (ZEN-ovalbumin (ZEN-OVA)) and developed an indirect competitive-ELISA (ic-ELISA) strategy using monoclonal antibodies (anti-ZEN-Abs) with high sensitivity and specificity. On this basis, a visual multicolor ELISA was developed based on alkaline phosphatase (AP) converting ascorbic acid–phosphate (VcP) to ascorbic acid (Vc) to control the growth of AuNRs. Compared with traditional ic-ELISA for ZEN, the visualized multicolor ELISA based on AuNRs growth offered a more convenient and intuitive strategy for the detection of ZEN contaminant (Scheme 1).

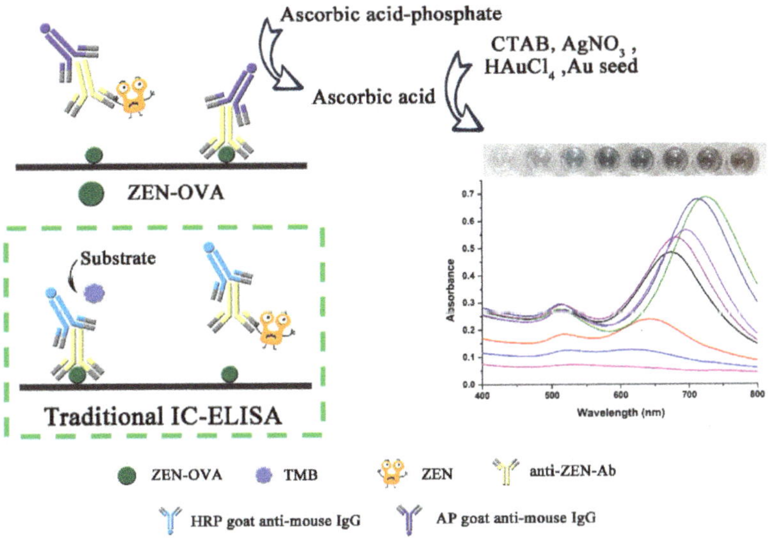

Scheme 1. Procedure of the visualized AuNRs growth-based multicolor ELISA for ZEN detection compared with the traditional ic-ELISA.

2. Materials and Methods

2.1. Material and Apparatus

Hexadecyl trimethyl ammonium bromide (CTAB, 99%) was purchased from Solarbio (Beijing, China). Chloroauric acid ($HAuCl_4$) for the synthesis of AuNRs was obtained from Sigma-Aldrich (St. Louis, MO, USA). Silver nitrate ($AgNO_3$), Vc, VcP, and sodium borohydride ($NaBH_4$) for the synthesis of AuNRs, *1*-ethyl-3-[*3*-(dimethylamino) propyl] carbodimide (EDC) and OVA for the synthesis of the conjugate of ZEN-OVA were obtained from the Sinopharm Chemical Reagent Co., Ltd. (Shanghai, China). Pyridine and carboxymethoxylamine (CMO) were purchased from TCI Development Co. Ltd. (Shanghai, China). The anti-ZEN monoclonal antibody (anti-ZEN-Ab, 1.0 mg/mL) was purchased from Shandong Lvdu Biotechnology Co. Ltd. (Shandong, China). *3,3,5,5*-tetramethylbenzidine (TMB), HRP goat anti-mouse IgG conjugated (1.0 mg/mL), AP goat anti-mouse IgG conjugated (3.5 mg/mL), AFB_1, ZEN and the structural analogues (α-zearalenol, α-zearalanol, β-zearalenol, zearalanone, β-zearalanol) (1.0 mg/L) were also purchased from Sigma-Aldrich (St. Louis, MO, USA). T-2 toxin, ochratoxin A (OTA), and fumonisin B_2 (FB_2) were purchased from Toronto Research Chemicals (Toronto, ON, Canada).

The microplate reader for reading the absorbance values was purchased from Thermo Fisher Scientific (Waltham, MA, USA). The UV-visible spectrophotometer (Cary 50 Bio) was obtained from Varian (Salt Lake, CA, USA). The 96-well polystyrene microplates, multichannel pipettes (100–300 μL), and single-channel pipettes (2.5–1000 μL) were obtained from Thermo Fisher Scientific (Waltham, MA, USA). The transmission electron microscopy (TEM) images were obtained from Talos G2 200X electronic microscope (Thermo Fisher Scientific (Waltham, MA, USA). A vortex machine (HQ-60) was purchased from North Tongzheng Biotechnology Development Company (Beijing, China). Milli-Q Ultrapure Water System was purchased from Milli-Q Millipore, (Bedford, MA, USA).

2.2. Preparation of the Coating Antigen—The Conjugate of ZEN-OVA

The preparation method of ZEN-CMO was modified according to the previous literature [30]. Briefly, 10.0 mg of ZEN was dissolved in 200 μL of methanol and mixed well. Then, 20.0 mg of CMO and 1.0 mL of anhydrous pyridine were sequentially added into the above mixed solution. The mixture was stirred and reacted under nitrogen protection for 24 h. After the reaction was completed, the mixture was placed in a vacuum oven at 80 °C to remove pyridine and store at 4 °C for later use. We mixed 9.78 mg of the above conjugate ZEN-CMO and 14.0 mg of EDC, successively dissolved it in 1.0 mL of DMF, and stirred and activated it overnight at 4 °C. In total, 10.0 mg of OVA was dissolved in 2.0 mL of $NaHCO_3$ (130 mmol/L) and precooled. The activated product was added dropwise under ice bath conditions. Two hours later, the product was placed under 4 °C overnight and dialyzed against PBS buffer solution for 72 h and stored at −20 °C.

2.3. Test Procedure of ic-ELISA

The antigen ZEN-OVA was dissolved in the coating solution and mixed evenly. The mixture was added to a microplate (0.025, 0.05, 0.1 μg/well, 100 μL/well), and incubated overnight at 4 °C. After washing the plate with PBST (0.01 mol/L PBS and 0.1% Tween-20) 3 times, the blocking solution (PBS containing 0.5% skimmed milk powder, 200 μL/well) was added, and we incubated the mixture at 37 °C for 1 h. After washing, the mixture of ZEN standards were diluted to different concentrations (50 μL) and anti-ZEN Ab (50 μL) was added and reacted at 37 °C for 1 h. Then, 100 μL of HRP goat anti-mouse IgG was added into each well and incubated for 0.5 h. After washing the plate with PBST 5 times, 100 μL of the TMB substrate solution was added into each well. After incubation for 15–30 min, the reaction was terminated with 50 μL of H_2SO_4 (1.25 mol/L). Then, the absorbance was measured at 450 nm using a microplate reader.

2.4. Specificity of ic-ELISA

The specificity of Abs is related to the structure of the antigenic determinant and is expressed as the cross-reactivity rate. Ten mycotoxins including ZEN, five structural analogs of ZEN (α-zearalenol, β-zearalenol, α-zearalanol, β-zearalanol, and zearalanone), and four common mycotoxins (AFB_1, OTA, T-2 toxin, and FB_2) were used as competing standards to determine the cross-reactivity rate to anti-ZEN-Ab by ic-ELISA. The mentioned analytes were diluted with PBS to the following concentrations (1000, 333.3, 111.11, 37.03, 12.34, 4.11, 1.37, 0.45, 0.15, 0.05, and 0.016 μg/L). The following equation was used to calculate the cross-reaction rate.

$$CR\ (\%) = IC_{50}\ (50\%\ \text{inhibitory concentration (ZEN)})/IC_{50}\ (50\%\ \text{inhibitory concentration (competitor)}) \times 100\% \tag{1}$$

2.5. AuNR Growth-Based Multicolor ELISA for ZEN

Preparation of AuNRs seed liquid. The AuNRs were synthesized by the seed-mediated method in this study. A total of 1.0 mL of CTAB (0.2 mol/L) and 1.0 mL of $HAuCl_4$ (0.5 mmol/L) were added into a round-bottom flask and mixed thoroughly. Then, 120 μL of $NaBH_4$ (0.01 mol/L) was added and stirred gently for 2 min. The obtained mixture was used as the seed liquid for AuNR growth.

Effect of Vc dosage on AuNR growth. The mixed solution containing 125 μL of CTAB (0.2 mol/L), 1.5 μL of $AgNO_3$ (0.01 mol/L), and 12.5 μL of $HAuCl_4$ (0.01 mol/L) was added to each well of the enzyme-labeled plate. After mixing, the Vc (0.01 mol/L) solution of different volumes (0–20 μL) was added to control the total volume of the solution to 240 μL. Next, 10 μL of the seed solution was added to mediate the production of AuNRs. After mixing evenly, the mixture was incubated at room temperature for 1 h, and UV absorption spectra of the grown AuNRs were recorded.

Procedure of visualized multicolor ELISA based on AuNR growth. The encapsulation process of the ZEN-OVA conjugate and its competitive binding with anti-ZEN Abs were the same as ic-ELISA. The difference was that 100 μL of AP-goat anti-mouse IgG diluted 10,000 times with Tris-HCl buffer (1 mmol/L pH 7.4) was added to each well. After incubation at 37 °C for 30 min and full washing, 80 μL of VcP (15 mmol/L) was added into each well. After full mixing, the mixture was incubated at 37 °C for 1 h. Then, 50 μL of the above reaction solution was mixed evenly with the solution containing 125 μL of CTAB (0.2 mol/L), 1.5 μL of $AgNO_3$ (0.01 mol/L), and 12.5 μL of $HAuCl_4$ (0.01 mol/L) and controlled to 240 μL with ultrapure water. After 10 μL of seed solution was added, the mixture was incubated at room temperature for 1 h and tested by naked eye and UV absorption spectroscopy.

3. Results and Discussion

3.1. Preparation and Characterization of ZEN-CMO and ZEN-OVA Conjugate

In this study, the prepared ZEN-CMO (Mr: 391.28) was analyzed by mass spectrometry (Figure S1). In an anionic environment, the characteristic ion peaks of ZEN-CMO $[M-1]^-$ at 390.40 and $[2M-1]^-$ at 781.80 were observed, proving the successful preparation of the hapten ZEN-CMO with acceptable purity. The ZEN-CMO product was further conjugated with OVA to obtain ZEN-OVA conjugate, the concentration of which was determined to be 3.85 mg/mL by a commercial BCA kit. Furthermore, the prepared ZEN-OVA conjugate was tested for its conjugate ability to anti-ZEN Abs (Table S1). When the Abs was diluted 16,000 times, the OD_{450} value reached 1.253, and the inhibition rate reached 98.92%, indicating the successful coupling of ZEN-CMO and OVA, and the obtained ZEN-OVA conjugate could be used for subsequent immunoassay experiments.

3.2. Conditions Optimization of Traditional ELISA

Dilution times of ZEN-OVA conjugate and anti-ZEN Abs. Different amounts of ZEN-OVA conjugate were coated on a 96-well microtiter plate to optimize the dilution times through ic-ELISA. According to the results shown in Table S2, when the coating amount of the ZEN-OVA conjugate was 0.1 μg/well and the Ab dilution factor was 64,000, the OD_{450} value reached 0.858, and the IC_{50} was 2.68 μg/L. When the coating amount was 0.05 μg/well and 0.025 μg/well, the dilution factor was 32,000, the corresponding OD_{450} and IC_{50} values were 1.078 and 0.793, 1.06 μg/L and 0.85 μg/L, respectively. It can be clearly seen that the IC_{50} increased with the increase in the coating amount of ZEN-OVA, which was because the excessive amount of ZEN-OVA bound to the Ab led to the decrease in detection sensitivity. When the coating amount was 0.025 μg/well and the Ab dilution factor was 32,000, the color development was relatively stable and the IC_{50} reached the lowest value, which was selected as the optimal condition for subsequent experiments.

Blocking Solution. The blocking solution was applied to block the excess binding sites in the micropores. A higher concentration of the blocking solution may affect the subsequent binding between the antigen and Ab, thereby reducing the sensitivity of the method. In the study, the IC_{50} values of the ic-ELISA method using different concentrations (0.5% and 1%) of skimmed milk powder as the blocking solution were compared (Table S3). The 0.5% skimmed milk powder had an IC_{50} of 0.85 μg/L, less than using 1% skimmed milk powder (2.41 μg/L), which was selected as the blocking solution.

pH value of PBS diluent. PBS buffers with different pH values (5.7, 7.4, and 8.5) were used for Ab dilution and ic-ELISA tests were performed (Table S4). When PBS was used at pH 7.4, the IC_{50} value was only 0.85 μg/L, which was significantly lower than that obtained at pH 5.7 and 8.5 (2.95 μg/L and 1.86 μg/L). At this time, the corresponding absorbance value (λ = 450 nm) was also significantly higher than the other two pH values, because the acidic or alkaline environment affected the binding reaction between the antigen and Ab as well as enzyme-labeled antibody, resulting in the reduction in detection sensitivity. Therefore, a PBS buffer of pH 7.4 was chosen as a diluent for Ab and ZEN standards.

3.3. ic-ELISA Standard Curve

Under the optimal conditions: coating amount of ZEN-OVA 0.025 μg/well, Ab dilution 32,000 times, blocking solution 0.5% skim milk powder in PBS, PBS buffer (pH 7.4) as diluent, the standard curve of ic-ELISA for ZEN is shown in Figure 1. The sensitivity (IC_{50}) and limit of detection (IC_{15}) reached 0.85 ± 0.04 μg/L and 0.22 ± 0.08 μg/L, indicating that this method can provide accurate and sensitive analysis for the ZEN toxin.

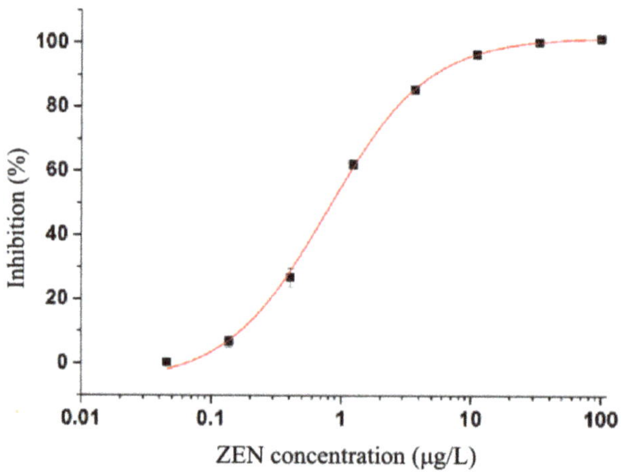

Figure 1. Standard curve of ZEN by ic-ELISA.

3.4. Specificity of Traditional ELISA

In order to evaluate the specificity of the established ic-ELISA method to ZEN, five ZEN structural analogs (α-zearalanol, β-zearalanol, α-zearalenol, β-zearalenol, and zearalenone) and four common mycotoxins (AFB_1, OTA, FB_2, and T-2 toxin) were selected and analyzed (Table 1). The cross-reaction rate of four structural analogs α-zearalanol, β-zearalanol, α-zearalenol, and β-zearalenol were 35.27%, 45.70%, 29.72% and 17.93%, respectively; the cross-reaction rate of zearalenone was only 1.58%; and there was no obvious cross-reaction with other mycotoxins. These results proved that the established ic-ELISA method had high specificity.

Table 1. Cross-reacting of ZEN with other mycotoxins.

Determinand	Structure	IC_{50} (μg/L)	Cross-Reaction Rate
Zearalenone	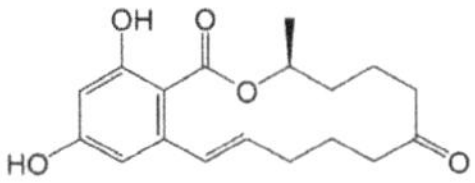	0.85	100
α-Zearalanol	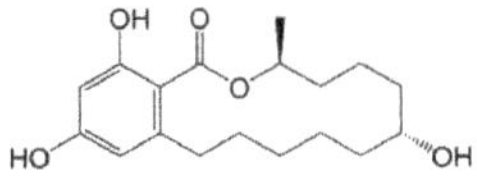	2.41	35.27
β-Zearalanol	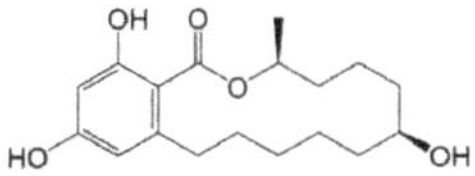	1.86	45.70
α-Zearalenol	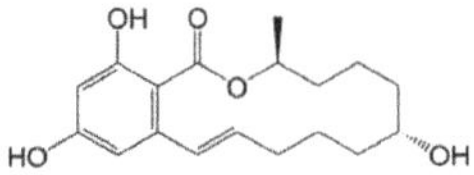	2.86	29.72
β-Zearalenol	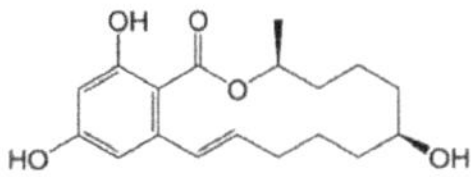	4.74	17.93
Zearalanone	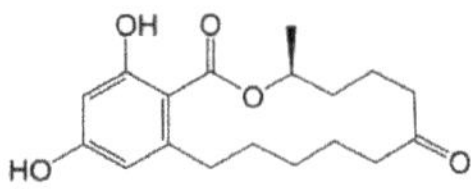	53.79	1.58
AFB_1	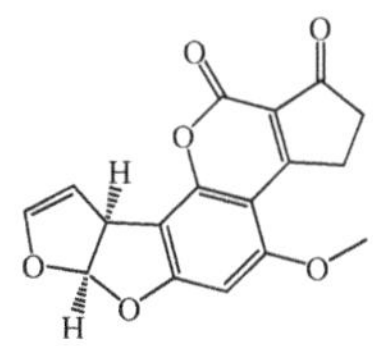	>1000	<0.01
OTA	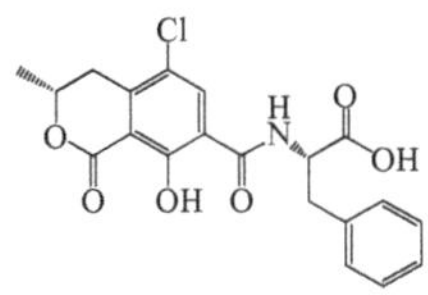	>1000	<0.01
FB_2	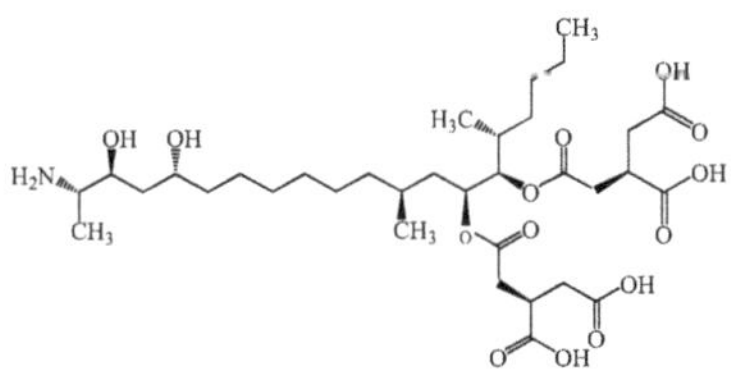	>1000	<0.01
T-2 toxin	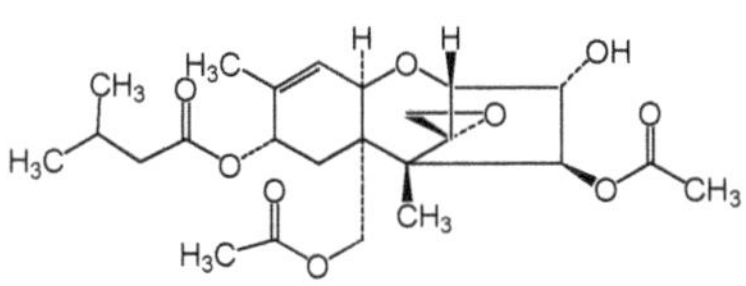	>1000	<0.01

3.5. Multicolor ELISA Based on AuNRs Growth

The AuNRs used in the study were prepared by a traditional seed-mediated method. The amount of Vc had a very important effect on the morphology and properties of the prepared AuNPs. Figure 2a shows the results of AuNRs obtained under different Vc solution additions. When a small amount of Vc solution was added (No. 1–6), the mixed solution appeared yellow, and the corresponding UV-visible absorption spectrum had a significant absorption peak at 400 nm, which was the characteristic absorption of $HAuCl_4$. With the increase in Vc solution volume, the color of the solution gradually lightened. When the added amount was 12.0 μL, the solution was colorless by naked eye observation. Meanwhile, no significant absorption was observed in the wavelength range greater than 500 nm. This was because when the amount of Vc was low, only Au(III) was reduced to colorless Au(I) (No. 7). When the amount of Vc further increased, a new and gradually enhanced UV absorption peak was formed at 500–800 nm, indicating that AuNRs gradually grew (No. 8–11), which also proved that the Vc supplemental level played a very important role in regulating AuNRs growth.

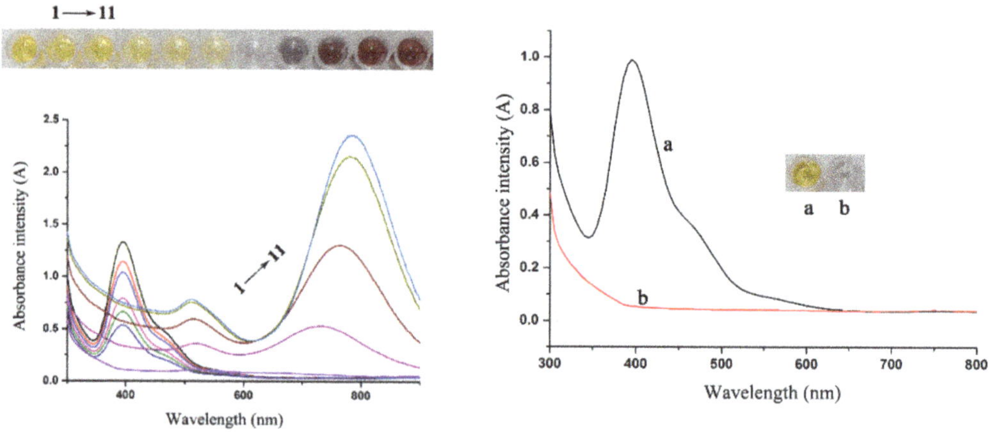

(a) (b)

Figure 2. (**a**) UV-visible absorption spectra of AuNRs grown with different volumes of Vc; (**b**) UV-vis spectra and visual results of the AuNRs growth solution before and after adding excessive VcP (a: before; b: after).

AP can induce the conversion of VcP to Vc and then promote the growth of AuNRs. Figure 2b shows the UV absorption spectra and visual results of the AuNR growth solution before and after the addition of excessive VcP. It was clearly observed that the addition of VcP changed the growth solution from yellow (curve a) to colorless (curve b). In addition, when excessive VcP was added, there was no absorption peak in the wavelength range of 500–800 nm, indicating that VcP could also reduce Au (III) to Au (I) but could not promote the growth of AuNRs. It was confirmed that the conversion of VcP to Vc by AP was a necessary prerequisite for promoting AuNR growth. Furthermore, different amounts of AP-labeled secondary Abs were mixed with a fixed amount of Vc sodium phosphate and incubated at a certain temperature to generate different amounts of Vc, which then produced AuNRs with different aspect ratios, colors, and UV absorption spectrum, which can be used as the basis for visual detection. Figure 3 shows the effect of adding different volumes of AP-goat anti-mouse IgG enzyme-labeled secondary Ab on the growth of AuNRs. With the increase in the amount of enzyme-labeled secondary Ab, the color of the mixed solution gradually became darker, from colorless to reddish brown. The UV absorption spectra showed that the longitudinal absorption peak of the prepared AuNRs had a significant red shift. There was a good linear relationship between the amount of AP goat anti-mouse IgG conjugated secondary Ab (40–80 μL) and the longitudinal absorption peak of the AuNRs produced (R^2 = 0.995) (Figure 3b).

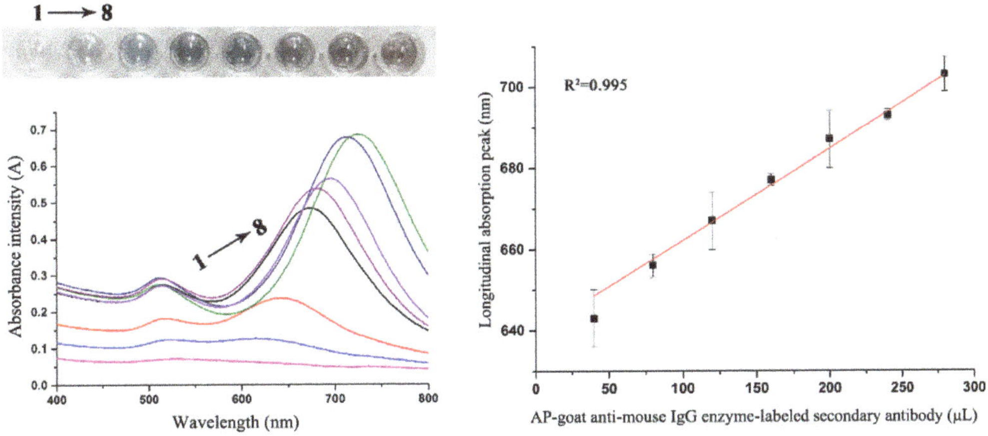

(a) (b)

Figure 3. (**a**) UV-visible absorption spectra of AuNRs grown with different volumes of AP-labeled secondary Ab; (**b**) the standard curve of AP-labeled secondary Ab and longitudinal absorption peaks of AuNRs.

Transmission electron micrographs (TEM) of AuNRs prepared with the addition of 120 and 280 μL of enzyme-labeled secondary Abs are shown in Figure 4. The size of the AuNRs prepared with 120 μL of enzyme-labeled secondary Ab were uniform and moderate, with an average aspect ratio of 3.5 (Figure 4a,b). When 280 μL of enzyme-labeled secondary Ab was used, the aspect ratio of AuNRs increased to 4.5 (Figure 4c,d). The results showed that the addition of AP goat anti-mouse IgG enzyme-labeled secondary Ab had a direct effect on the increase in the AuNRs aspect ratio. The color and UV-visible absorption spectrum of the mixed solution showed a linear relationship with the content of AP goat anti-mouse IgG enzyme-labeled secondary Ab, indicating the feasibility of developing a visual multicolor ELISA.

3.6. Standard Curve of AuNPs Growth-Based Multicolor ELISA for ZEN Detection

The results of multicolor ELISA based on AuNR growth for ZEN at different concentrations (0–200 μg/L) are illustrated in Figure 5. The mixed solution without the addition of ZEN standard had a distinct brownish red color. As the content of ZEN increased, the color of the mixture became lighter, from dark gray to green to colorless. When the ZEN concentration was 0.1 μg/L, the mixed solution showed a color change that can be recognized by the naked eye (from brownish red to dark gray), which meant the limit of detection (LOD) of the developed AuNRs growth-based multicolor ELISA for ZEN was lower than 0.1 μg/L. The UV absorption spectra showed that the longitudinal absorption peak of AuNRs shifted to blue gradually with the increase in ZEN concentration. (Figure 5a). As shown in Figure 5b, ZEN concentration had a good linear correspondence with the longitudinal absorption wavelength of the corresponding AuNRs in the range of 0.001–100 μg/L, which strongly demonstrated the potential applicability of the developed visual multicolor ELISA based on AuNRs growth for naked identification of low concentrations of mycotoxins.

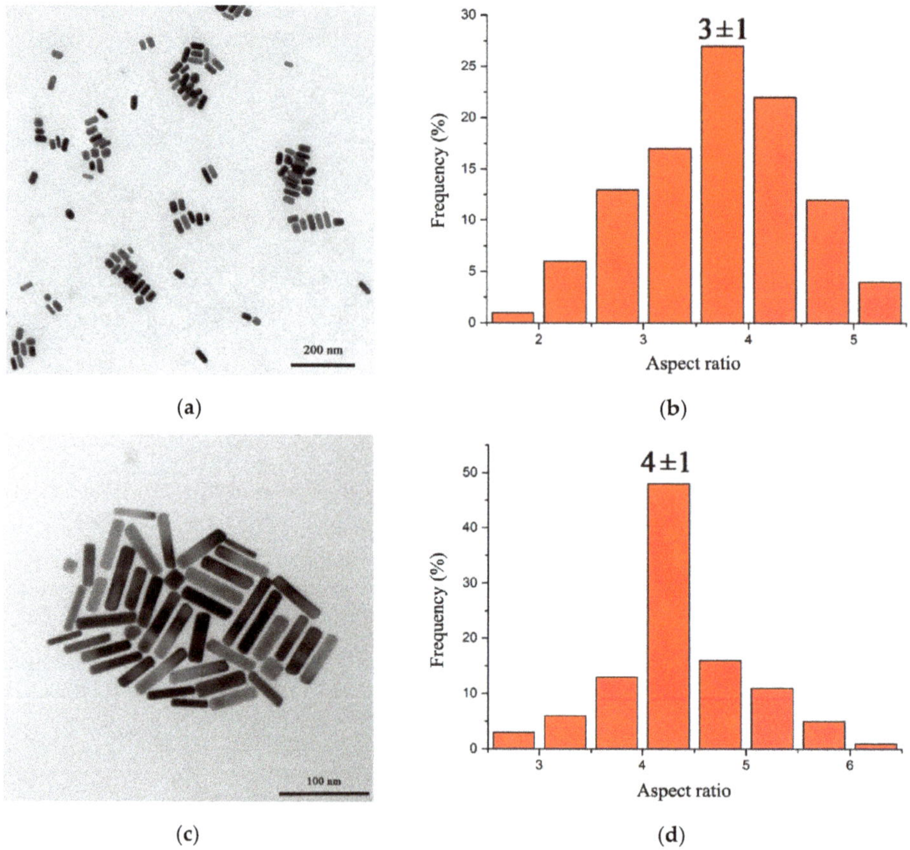

Figure 4. TEM images and aspect ratio of particle size analysis of AuNRs prepared by adding 120 (**a**,**b**) and 280 μL (**c**,**d**) of enzyme-labeled secondary Abs.

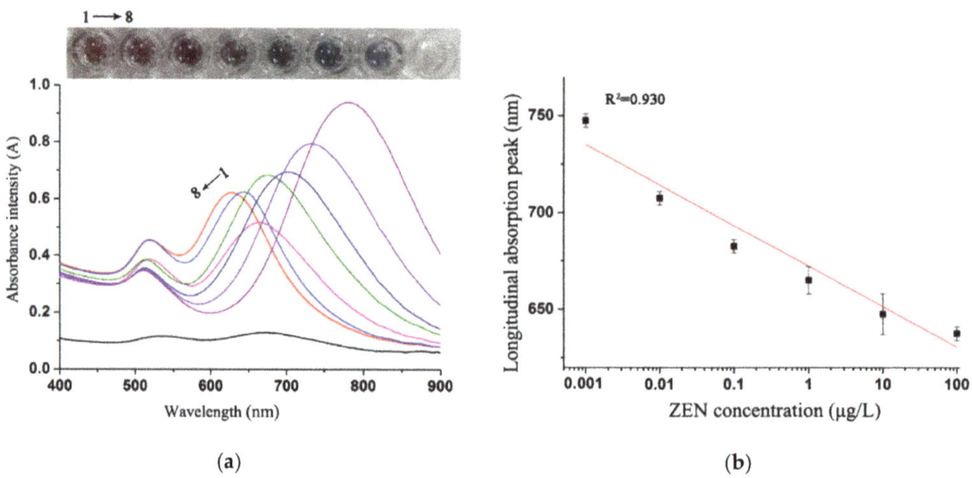

Figure 5. (**a**) UV absorption spectra and visual results of AuNR growth-based multicolor ELISA; (**b**) Standard curve of different ZEN concentration and AuNR longitudinal absorption peak.

4. Conclusions

This study developed two detection strategies for the ZEN toxin based on the specific reaction of antigen–antibody. The established ic-ELISA method had high accuracy, sensitivity, and specificity, and can provide an effective semiquantitative analysis strategy for ZEN contaminants in food. The visual analysis strategy based on the growth of AuNRs was suitable for the rapid screening of a large number of contaminated samples ZEN and can be extended to the monitoring and control of the content of other harmful substances in foods. In Table 2, various reported ZEN detection strategies are illustrated to demonstrate the merits of the developed two methods in this study.

Table 2. Comparison of the published traditional instrumental and immunoassays for ZEN detection.

Method Applied	Materials	Limit of Detection	References
GC-MS	-	5 ng/g	[15]
LC-MS/MS	-	0.02 ng/mL	[18]
Lateral flow immunochromatographic assays	Colloidal gold, quantum dots, polystyrene microspheres	10, 1, 1 μg/L	[22]
Fluorescence-linked immunosorbent assay	CdTe/CdS/ZnS quantum dots	0.012 ng/mL	[27]
Multiplexed immunochromatographic assay	Dual CdSe/ZnS quantum dot nanobeads	10 ng/mL	[31]
ic-ELISA	-	0.22 ± 0.08 μg/L (IC_{15})	This work
Visualized multicolor ELISA	AuNRs	0.1 μg/L	This work

Supplementary Materials: The following are available online at https://www.mdpi.com/article/10.3390/foods10112654/s1, Figure S1: Mass spectrum of ZEN-CMO, Table S1: Results of ZEN-OVA binding to Abs, Table S2: Optimization results of the coating-antigen and Ab, Table S3: Optimization results of the blocking buffer, and Table S4: Optimization results of pH value of PBS buffer.

Author Contributions: T.M. and K.L. completed the experiments and organized the writing of the entire manuscript (T.M. and K.L. contributed to this article equally); X.Y. checked the language and format of the manuscript; J.Y. participated in some experiments. M.P. provided the idea and financial support for the research and completed Section 4. S.W. checked the quality of the article. All authors have read and agreed to the published version of the manuscript.

Funding: This research was funded by the National Key Research and Development Program of China (No. 2017YFC1600402), the National Natural Science Foundation of China (No. 31972147), the Open Project Program of State Key Laboratory of Food Nutrition and Safety, Tianjin University of Science and Technology (No. SKLFNS-KF-202115), and the Natural Science Foundation of the Inner Mongolia Autonomous Region (No. 2021MS03077).

Institutional Review Board Statement: Not applicable.

Informed Consent Statement: Not applicable.

Data Availability Statement: The datasets generated for this study are available on request to the corresponding author.

Conflicts of Interest: The authors declare no conflict of interest.

References

1. Freire, L.; Sant'Ana, A.S. Modified mycotoxins: An updated review on their formation, detection, occurrence, and toxic effects. *Food Chem. Toxicol.* **2018**, *111*, 189–205. [CrossRef] [PubMed]
2. Mahato, D.K.; Devi, S.; Pandhi, S.; Sharma, B.; Maurya, K.K.; Mishra, S.; Dhawan, K.; Selvakumar, R.; Kamle, M.; Mishra, A.K.; et al. Occurrence, impact on agriculture, human health, and management strategies of zearalenone in food and feed: A review. *Toxins* **2021**, *13*, 92. [CrossRef]
3. Mazaheri, M.; Maymand, M.M.; Gilasgar, A.; Akbarzadeh, A.; Manafi, M.H. Quantification of the zearalenone in maize oil with no clean-up. *Food Control* **2021**, *127*, 108166. [CrossRef]

4. Zinedine, A.; Soriano, J.M.; Molto, J.C.; Manes, J. Review on the toxicity, occurrence, metabolism, detoxification, regulations and intake of zearalenone: An oestrogenic mycotoxin. *Food Chem. Toxicol.* **2007**, *45*, 1–18. [CrossRef]
5. Yu, Z.; Zhang, L.; Wu, D.; Liu, F. Anti-apoptotic action of zearalenone in MCF7 cells. *Ecotox. Environ. Saf.* **2005**, *62*, 441–446. [CrossRef]
6. Rai, A.; Das, M.; Tripathi, A. Occurrence and toxicity of a fusarium mycotoxin, zearalenone. *Crit. Rev. Food Sci.* **2020**, *60*, 2710–2729. [CrossRef]
7. Sneha, S.; Baker, S.C.; Green, A.; Storr, S.; Aiyappa, R.; Martin, S.; Pors, K. Intratumoural Cytochrome P450 expression in breast cancer: Impact on standard of care treatment and new efforts to develop tumour-selective therapies. *Biomedicines* **2021**, *9*, 290. [CrossRef] [PubMed]
8. Yu, Z.L.; Hu, D.; Li, Y. Effects of zearalenone on mRNA expression and activity of cytochrorne P450 1A1 and 1B1 in MCF-7 cells. *Ecotox. Environ. Safe.* **2004**, *58*, 187–193. [CrossRef]
9. Rogowska, A.; Pomastowski, P.; Sagandykova, G.; Buszewski, B. Zearalenone and its metabolites: Effect on human health, metabolism and neutralisation methods. *Toxicon* **2019**, *162*, 46–56. [CrossRef] [PubMed]
10. Gajęcka, M.; Janowski, T.; Jakimiuk, E.; Zielonka, Ł.; Podhalicz-Dziegielewska, M.; Jan Obremski, K.; Gajęcki, M. Influence of long-term zearalenone intoxication on the concentration of progesterone and 17 beta-oestradiol in blood plasma in bitches. *Bull. Vet. Inst. Pulawy* **2008**, *52*, 405–409.
11. Ji, F.; He, D.; Olaniran, A.O.; Mokoena, M.P.; Xu, J.; Shi, J. Occurrence, toxicity, production and detection of Fusarium mycotoxin: A review. *Food Prod. Process. Nutr.* **2019**, *1*, 6. [CrossRef]
12. Hong, X.; Mao, Y.H.; Yang, C.Q.; Liu, Z.J.; Li, M.; Du, D.L. Contamination of Zearalenone from China in 2019 by a Visual and Digitized Immunochromatographic Assay. *Toxins* **2020**, *12*, 521. [CrossRef]
13. Pan, M.F.; Ma, T.Y.; Yang, J.Y.; Li, S.J.; Liu, S.M.; Wang, S. Development of lateral flow immunochromatographic assays using colloidal au sphere and nanorods as signal marker for the determination of zearalenone in cereals. *Foods* **2020**, *9*, 281. [CrossRef]
14. GB2761-2017. National Standard of People's Republic of China-Limitation of Mycotoxins in Food. Available online: http://down.foodmate.net/info/sort/2/7407.html (accessed on 17 September 2017).
15. Tanaka, T.; Yoneda, A.; Inoue, S.; Sugiura, Y.; Ueno, Y. Simultaneous determination of trichothecene mycotoxins and zearalenone in cereals by gas chromatography-mass spectrometry. *J. Chromatogr. A* **2000**, *882*, 23–28. [CrossRef]
16. Amelin, V.G.; Karaseva, N.M.; Tretyakov, A.V. Simultaneous determination of trichothecene micotoxins, ochratoxin A, and zearalenone in grain and products of its processing, feed premixes, and meat by gas chromatography. *J. Anal. Chem.* **2013**, *68*, 61–67. [CrossRef]
17. Qian, M.R.; Zhang, H.; Wu, L.Q.; Jin, N.; Wang, J.M.; Jiang, K.Z. Simultaneous determination of zearalenone and its derivatives in edible vegetable oil by gel permeation chromatography and gas chromatography-triple quadrupole mass spectrometry. *Food Chem.* **2015**, *166*, 23–28. [CrossRef] [PubMed]
18. Li, C.L.; Deng, C.L.; Zhou, S.; Zhao, Y.F.; Wang, D.; Wang, X.D.; Gong, Y.Y.; Wu, Y.N. High-throughput and sensitive determination of urinary zearalenone and metabolites by UPLC-MS/MS and its application to a human exposure study. *Anal. Bioanal. Chem.* **2018**, *410*, 5301–5312. [CrossRef] [PubMed]
19. Sun, D.L.; Li, C.L.; Zhou, S.; Zhao, Y.F.; Gong, Y.Y.; Gong, Z.Y.; Wu, Y.N. Determination of trace zearalenone and its metabolites in human serum by a high-throughput UPLC-MS/MS analysis. *Appl. Sci.* **2019**, *9*, 741. [CrossRef]
20. Tahoun, I.F.; Gab-Allah, M.A.; Yamani, R.N.; Shehata, A.B. Development and validation of a reliable LC-MS/MS method for simultaneous determination of deoxynivalenol and T-2 toxin in maize and oats. *Microchem. J.* **2021**, *169*, 106599. [CrossRef]
21. Engvall, E. Enzyme-linked immunosorbent assay (ELISA), quantitative assay of immunoglobulin G. *Immunochemistry* **1971**, *8*, 871–874. [CrossRef]
22. Li, S.J.; Sheng, W.; Wen, W.J.; Gu, Y.; Wang, J.P.; Wang, S. Three kinds of lateral flow immunochromatographic assays based on the use of nanoparticle labels for fluorometric determination of zearalenone. *Microchim. Acta* **2018**, *185*, 238. [CrossRef]
23. Gemes, B.; Takacs, E.; Gadoros, P.; Barocsi, A.; Kocsanyi, L.; Lenk, S.; Csakanyi, A.; Kautny, S.; Domjan, L.; Szarvas, G.; et al. Development of an immunofluorescence assay module for determination of the mycotoxin zearalenone in water. *Toxins* **2021**, *13*, 182. [CrossRef]
24. Li, R.X.; Wen, Y.; Yang, L.Q.; Liu, A.G.; Wang, F.L.; He, P.L. Dual quantum dot nanobeads-based fluorescence-linked immunosorbent assay for simultaneous detection of aflatoxin B1 and zearalenone in feedstuffs. *Food Chem.* **2022**, *366*, 130527. [CrossRef]
25. Liu, R.X.; Shi, R.R.; Zou, W.T.; Chen, W.H.; Yin, X.C.; Zhao, F.C.; Yang, Z.Y. Highly sensitive phage-magnetic-chemiluminescent enzyme immunoassay for determination of zearalenone. *Food Chem.* **2020**, *325*, 126905. [CrossRef]
26. Fu, J.M.; Zhou, Y.F.; Huang, X.L.; Zhang, W.J.; Wu, Y.H.; Fang, H.; Zhang, C.Z.; Xiong, Y.H. Dramatically enhanced immunochromatographic assay using cascade signal amplification for ultrasensitive detection of Escherichia coli O157: H7 in milk. *J. Agric. Food Chem.* **2020**, *68*, 1118–1125. [CrossRef] [PubMed]
27. Zhang, F.Y.; Liu, B.; Sheng, W.; Zhang, Y.; Liu, Q.; Li, S.J.; Wang, S. Fluoroimmunoassays for the detection of zearalenone in maize using CdTe/CdS/ZnS quantum dots. *Food Chem.* **2018**, *255*, 421–428. [CrossRef] [PubMed]
28. Xiong, Y.; Pei, K.; Wu, Y.Q.; Duan, H.; Lai, W.H.; Xiong, Y.H. Plasmonic ELISA based on enzyme-assisted etching of Au nanorods for the highly sensitive detection of aflatoxin B-1 in corn samples. *Sens. Actuators B Chem.* **2018**, *267*, 320–327. [CrossRef]

29. Liu, Z.J.; Wang, X.Y.; Ren, X.X.; Li, W.B.; Sun, J.F.; Wang, X.W.; Huang, Y.Q.; Guo, Y.G.; Zeng, H.W. Novel fluorescence immunoassay for the detection of zearalenone using HRP-mediated fluorescence quenching of gold-silver bimetallic nanoclusters. *Food Chem.* **2021**, *355*, 129633. [CrossRef]
30. Thouvenot, D.; Morfin, R.F. Radioimmunoassay for zearalenone and zearalanol in human serum: Production, properties, and use of porcine antibodies. *Appl. Environ. Microb.* **1983**, *45*, 16–23. [CrossRef] [PubMed]
31. Duan, H.; Li, Y.; Shao, Y.N.; Huang, X.L.; Xiong, Y.H. Multicolor quantum dot nanobeads for simultaneous multiplex immunochromatographic detection of mycotoxins in maize. *Sens. Actuators B Chem.* **2019**, *291*, 411–417. [CrossRef]

foods

Article

Detection of Pyrethroids in Food by Immunofluorescence Enhanced Method Based on Three-Layer Core-Shell Structure Upconversion Materials

Lingyan Zhao [1], Jingyi Jin [1], Wenbo Zhu [1], Yuehua Zuo [2] and Yang Song [1,3,*]

[1] Tianjin Key Laboratory of Animal and Plant Resistance, Tianjin Normal University, Tianjin 300387, China; 2010170003@stu.tjnu.edu.cn (L.Z.); 2110170005@stu.tjnu.edu.cn (J.J.); 1910170006@stu.tjnu.edu.cn (W.Z.)
[2] Technical Center for Safety of Industrial Products of Tianjin Customs District, Tianjin 300308, China; zuoyuehua1987@126.com
[3] State Key Laboratory of Food Nutrition and Safety, Tianjin University of Science and Technology, Tianjin 300457, China
* Correspondence: skysongy@mail.tjnu.edu.cn; Tel.: +86-22-23765561

Abstract: A novel rare earth upconversion nanomaterial with a three-layer sandwich core–shell structure was synthesized by an improved thermal decomposition method, and the morphology, fluorescence intensity and diffraction peak position of the new material were characterized by TEM (transmission electron microscope), XRD (Powder X-ray diffraction)and fluorescence spectrophotometer. The inert core/active shell/inert shell design improved the upconversion luminous efficiency of the new material several times. FTIR (Fourier transform infrared spectroscopy)characterization showed that the surface of activated upconversion nanoparticles was modified with silicon shell and amino group. Combined with the characteristics that aminoated polystyrene magnetic microspheres could be separated by the magnetic field, an upconversion magnetic separation immunoassay method for the detection of pyrethroid pesticide residues was established. The capture probe competed with the pyrethroid standard, combined the signal probe, and measured the fluorescence signal value formed by the capture probe signal probe complex at 542 nm under 980 nm excitation light. The LOD (limit of detection)of fenpropathrin was 0.01 μg/L, cypermethrin was 0.015 μg/L, and fenvalerate was 0.011 μg/L. Through the actual sample detection of apple, cabbage and other samples, the recovery rate of pyrethroids was between 83.4~97.8%. The comparison with the HPLC (high performance liquid chromatography)detection results showed that the established method had good accuracy, and could realize the quantitative analysis of pyrethroids in food.

Keywords: pyrethroids; upconversion nanomaterials; fluorescence immunoassay

Citation: Zhao, L.; Jin, J.; Zhu, W.; Zuo, Y.; Song, Y. Detection of Pyrethroids in Food by Immunofluorescence Enhanced Method Based on Three-Layer Core-Shell Structure Upconversion Materials. *Foods* **2022**, *11*, 990. https://doi.org/10.3390/foods11070990

Academic Editor: Elena González-Peñas

Received: 9 March 2022
Accepted: 25 March 2022
Published: 29 March 2022

1. Introduction

Upconversion nanomaterials (UCNPs) follow the anti-stoke luminescence law, which could convert long wave low-energy photons into short wave high-energy photons [1,2], and have stable photochemical properties. Rare earth element doped upconversion materials could not only emit strong visible light under the excitation of near-infrared but also have the advantages of small light damage, high penetration depth, low spontaneous fluorescence interference of biological tissue and low biological toxicity [3–6]. Therefore, they have attracted extensive attention in the fields of medical imaging, detection and sensing. UCNPs could be combined with fiber chromatography materials to prepare test strips to detect veterinary drug residues in aquatic products and livestock and poultry meat. They could also be used to develop quantum dot biosensors to detect heavy metal ions in food, which shows that UCNPs have broad application prospects in food safety detection. However, the defects of low luminous efficiency and short fluorescence life have limited the research and application of UCNPs. Therefore, researchers at home and abroad have prepared upconversion materials with core–shell structures to improve their optical properties. For example, in 2020, Wu Haoyu prepared inert core–shell structure $NaYF_4$:Yb/Er@$NaGdF_4$, and compared with $NaYF_4$:Yb/Er without shell, its luminous efficiency was increased by more than 20 times [7]. In the same year, Zhu Yuyan prepared the active core–shell structure $NaYbF_4$@$NaYF_4$:Yb/Er, and the surface was coated with $NaYbF_4$ shell doped with sensitizer Yb^{3+}; compared with shell-free nanoparticles $NaYF_4$:Yb/Er, the fluorescence intensity was increased by 6 times [8]. In addition to the double-layer core–shell structure, Qiu et al. prepared the $NaYF_4$:Yb^{3+}/Tm^{3+}@$NaYbF_4$@/$NaYF_4$ multilayer core–shell structure of active core/active shell/inert shell, which increased the fluorescence brightness by 11 times compared with $NaYbF_4$@/$NaYF_4$ [9]. The research and development of multi-layer core–shell structure had created space confinement, doped rare earth ions in different shells, improved the upconversion efficiency and reduced the occurrence of cross relaxation [10].

Pyrethroids have been mainly used for the control of vegetable and fruit pests in agricultural planting and have become the second most common insecticide in agricultural production, accounting for about 30% of the pesticide market [11–14]. The latest national standard GB 2763-2021 stipulates that the maximum residue limit of fenpropathrin in fruit is 5 mg/kg, that in cabbage iss 1 mg/kg and that in cucumber is 0.2 mg/kg. The maximum residue limit of cypermethrin in apple, pear and cabbage is 2 mg/kg, and the maximum residue limit of fenvalerate in apple, pear and cabbage is 1 mg/kg; EU standard EU441/2012 stipulates that the maximum residue of bifenthrin in tomato is 0.3 mg/kg, and the maximum residues of fenvalerate and beta cypermethrin in potato are 0.02 mg/kg and 0.01 mg/kg [15–17]. Even though there are limited requirements in the national standards, the excessive residue of pyrethroid pesticides in edible agricultural products has been banned repeatedly. According to the national sampling data, pyrethroid pesticides are often used in leek, celery, green pepper, cowpea, pear and citrus, and the problem of unqualified rate was more prominent [18–20].

In this experiment, three-layer sandwich structure rare earth UCNPs were synthesized by the classical thermal decomposition method. The silicon shell and amino group were modified on the surface of UCNPs by the Stober method to increase hydrophilicity and biocompatibility. Based on the principle of upconversion magnetic separation immunofluorescence, an immunoassay method for rapid detection of pyrethroid residues in food was proposed. The additional amount of antigen and antibody in the detection system was optimized through bicinchonininc acid (BCA)kit to determine the optimal reaction conditions; to realize the rapid determination of pyrethroids in apple, pear, cabbage and cucumber samples, the accuracy of the method was verified by HPLC comparison experiment.

2. Materials and Methods

2.1. Materials and Instruments

Fenpropathrin, cypermethrin and fenvalerate were purchased from Adamas company (Shanghai, China). Enzyme labeled secondary antibody (Sheep anti rabbit) was purchased from Shanghai TITAN Technology Co., Ltd. (Beijing, China). Pyrethroid universal antigen, pyrethroid universal antibody, $YCl_3 \cdot 6H_2O$, $YbCl_3 \cdot 6H_2O$, $ErCl_3 \cdot 6H_2O$ were purchased from Bioengineering Shanghai Co., Ltd. (Shanghai, China). 1-octadecene, ammonium fluoride and 3-aminopropyl triethoxysilane (APTES, 98%) were purchased from Tianjin Hengshan Chemical Technology Co., Ltd. (Tianjin, China). Sodium hydroxide, methanol, absolute ethanol, acetonitrile and sodium chloride were purchased from Tianjin Chemical Reagent Supply and marketing company (Tianjin, China). Oleic acid, tetraethoxysilane (TEOS, 98%), CO-520 and cyclohexane were purchased from Tianjin Baiaotai Technology Development Co., Ltd. (Tianjin, China). Bovine serum albumin (BSA, 98%) was purchased from Merck, Germany. Glutaraldehyde was purchased from Shanghai Meiruier Chemical Technology Co., Ltd. (Shanghai, China). The ammonia and BCA protein concentration determination kit was purchased from Tianjin Yuanli Chemical Co., Ltd. (Tianjin, China). Aminoated polystyrene magnetic microspheres (MPM) were purchased from Beijing Beisile Color Co.,

Ltd. (Tianjin, China). The nanoparticle morphology and size of UCNPs were determined by FEI TECNAI G20 transmission electron microscope (TEM, FEI Company, Hillsboro, OR, USA). Fourier transform infrared spectroscopy (FTIR) of UCNPs was determined by FTIR spectrophotometer (Perkin Elmer, Waltham, MA, USA). Powder X-ray diffraction (XRD) measurements of UCNPs were performed by the AXIS-ULTRA-DLD instrument (Millipore, New York, NY, USA). All the above instruments were from the school of the chemistry of Tianjin Normal University (Tianjin, China). The fluorescence intensity of upconversion nanoparticles was detected by an f-2500 fluorescence spectrophotometer equipped with a 980 nm laser exciter. Waters e2695 HPLC was from Thermo Fisher Scientific, Waltham, MA, USA. All chemicals used were analytical grade.

2.2. Preparation and Surface Modification of UCNPs

Based on the high-temperature thermal decomposition method, 2 mmol $YCl_3 \cdot 6H_2O$, 30 mL1-octadecene and 12 mL oleic acid were put into a three-mouth bottle and heated to 170 °C under high-purity argon to form a yellow transparent solution. When the reactant was cooled to room temperature, methanol solution dissolved with 5 mmol NaOH and 8 mmol NH_4F was added drop by drop, stored at room temperature for 30 min, slowly raised the temperature to 100 °C, and methanol and water was evaporated in the reaction system. Following this, the temperature was raised again to 300 °C and kept for 1.5 h. After the reaction system was cooled to room temperature, the sample was centrifuged at 8500 rpm, washed several times alternately with ethanol and cyclohexane, dried in the oven and the $NaYF_4$ core was collected.

In additional sample preparation, 2 mmol $RECl_3 \cdot 6H_2O$ (RE:78% mol Y^{3+}, 2% mol Er^{3+}, 20% mol Yb^{3+}), 12 mL oleic acid and 30 mL octadecene was placed into a three-mouth bottle, raised the temperature to 170 °C under argon to form a yellow transparent solution, and cooled to room temperature. Next was the addition of 2 mmol $NaYF_4$ core dissolved in cyclohexane dropwise and stirred continuously, and the addition of methanol solution dissolved with NaOH and NH_4F and stirred continuously at room temperature. The reactant was heated to 100 °C and cyclohexane was evaporated. After the removal of cyclohexane, the procedure was the same as above. After being cleaned several times, it was dried in the oven and collected $NaYF_4$@$NaYF_4$:Yb,Er.

In addition, the same amount of $YCl_3 \cdot 6H_2O$, octadecene and oleic acid was placed into a three mouth bottle and heated to 170 °C to form a yellow transparent solution. This was then cooled to room temperature, to which was added 2 mmol $NaYF_4$@$NaYF_4$:Yb,Er dissolved in cyclohexane dropwise and stirred slowly. After cyclohexane removal, nucleation and growth, washing, drying and collection, the $NaYF_4$@$NaYF_4$:Yb, Er@$NaYF_4$ nanoparticles with the three-layer sandwich structure of inert core/active shell/inert core were obtained.

Finally, 100 mg UCNPs was weighed out, dispersed in 50 mL ethanol by ultrasonication and stirred quickly. Following this, 4 mL ammonia and 30mL secondary water was added and underwent vigorous stirring at 45°C. Four microliter TEOS was added to the mixture within 5 h. And one hundred microliter APTES solution was added dropwise to the suspension and reacted for 5 h. The sample was washed twice with ethanol and centrifuged to obtain white solid precipitation. This was then placed in the oven and dried for 24 h, after which UCNPs@SiO_2@NH_2 was collected and stored in dark condition prior to testing

2.3. Preparation of Capture Probe and Signal Probe

The signal probe was prepared by the glutaraldehyde crosslinking method. Here, 10 mg UCNPs@SiO_2@NH_2 was dissolved in 5 mL PBS (10 mmol/L), ultrasonicated for 20 min, 1.25 mL of 25% glutaraldehyde solution and 100 mg $NaBH_4$ was added drop by drop, shaken slowly for 1 h, centrifuged to remove the supernatant, washed with PBS for three times, dispersed the precipitate in 5 mL PBS, ultrasonicated for 10 min, added a certain amount of antibody, reacted for 6 h, and centrifuged to collect the supernatant and precipitate respectively. Then the precipitate was dissolved in 5 mL PBS (including 1% BSA), reacted for 6 h, and the precipitate was collected by centrifugation, dispersed in 5 mL PBS and stored at 4 °C to obtain the signal probe with the concentration of 2 mg/mL. When preparing the capture probe, the same amount of amino magnetic microspheres was dissolved in 5 mL PBS. The operation was the same as above. Following the addition of a certain amount of antigen, and centrifugation, the capture probe with a concentration of 2 mg/mL was obtained and stored at 4 °C.

In the process of preparing the two probes, the addition amount of pyrethroid antigen and antibody was optimized. Addition of 20 μg, 40 μg, 60 μg, 80 μg, 100 μg antigen and antibody respectively was done in preparing capture probes and signal probes. Calculation was done of the coupling amount through the BCA kit, and then the coupling rate was calculated according to the following formula to determine the optimal addition amount of antibody (antigen).

$$\text{coupling rate} = \frac{\text{Mc}}{\text{Mt}} \times 100\% \tag{1}$$

In the formula, Mc is the coupling amount of antibody (antigen), and Mt is the total amount of antibody (antigen).

2.4. Optimization of Capture Probe Addition

To ensure that the capture probe and signal probe reacted fully, the addition amount of capture probe was optimized in this study. When the addition amount of signal probe was determined, addition of 40 μg, 60 μg, 80 μg, 100 μg, 120 μg, 140 μg capture probes respectively, and the fluorescence value of the signal probe capture probe complex was measured.

2.5. Establishment of UpConversion Magnetic Separation Immunofluorescence Method

The pyrethroid standard competed with the capture probe to bind the signal probe. The capture probe signal probe conjugate was separated by the external magnetic field, and the upconversion magnetic separation immunofluorescence method was established. The pyrethroid standard competed with the capture probe to bind the signal probe. The capture probe signal probe conjugate was separated by the external magnetic field, and the upconversion magnetic separation immunofluorescence method was established. The standard solution was taken of fenpropathrin, cypermethrin and fenvalerate in different concentrations and mixed with the capture probe, added the signal probe, reacted at 25 °C for 1 h, separated the capture probe signal probe complex with the external magnetic field, collected the precipitation, washed it with PBS for three times, dispersed the precipitation in 2 mL PBS, and measured the fluorescence value with an external 980 nm excitation light source fluorescence spectrophotometer. The upconversion immunofluorescence standard curve at different concentrations was established.

2.6. Sample Pretreatment

Apple, pear, cabbage and cucumber were selected as experimental samples and respectively smashed in a blender. Following this, 2 g apple and pear samples were placed into a centrifuge tube, 10 mL acetone-n-hexane (1:1) mixed solvent was added, and subjected to vortexed oscillation for 10 s and ultrasonic extraction for 30 min. Then, sample was centrifuged at 4000 r/min for 5 min, and the supernatant was passed through a 0.22 μm micro membrane and stored for the test.

Two grams of cabbage and cucumber samples were prepared and placed in a centrifuge tube, to which 4 mL 0.1% acetic acid acetonitrile solution was added, and subjected to vortexed oscillation for 5 min; following this, 1 g sodium chloride was added, and subjected to vortexed oscillation for 2 min, centrifuged for 2 min at 4500 r/min, and 2 mL of supernatant was extracted. Finally, the sample was spun steamed and dried, fixed to a volume of 1 mL with methanol, passed through 0.22 μm microfilm, and stored for testing.

2.7. HPLC Comparison Test

2.7.1. Chromatographic Conditions

Chromatographic conditions were as follows: Chromatographic column: Bridge C18 (5 μm, 4.6 × 250 mm); Mobile phase: methanol–water (32:68); Flow rate: 1.0 mL/min; Column temperature: 30 °C; Injection volume: 20 μL; Detection wavelength: 205 nm.

2.7.2. Extraction and Purification

Samples were taken according to Appendix A of GB 2763-2021 [21]. Stems were removed from apple, pear and cucumber, and the root was removed from cabbage as the sample determination part. The samples were chopped and mixed into homogenate, sealed and frozen at −20 °C. Samples were extracted with acetonitrile, salted out and centrifuged with sodium chloride; the supernatant was purified by solid-phase extraction column, eluted with acetonitrile toluene solution (3 + 1), spun in the water bath until it was nearly dry, volume was fixed methanol, and it was filtered through 0.22 μm filter membrane, in preparation for test [22].

2.8. Actual Sample Testing

Three samples of apple, pear, Chinese cabbage and cucumber were randomly selected in the market, and the surface attachments were gently wiped with gauze. No pyrethroid pesticides were detected in 12 samples by HPLC. Three standards of fenpropathrin, cypermethrin and fenvalerate were randomly added into 12 blank samples. After the sequence was disrupted, they were renumbered 1–12#, tested simultaneously by this experimental method and HPLC, and the data were recorded.

2.9. Method Specificity

Fenpropathrin, Cypermethrin, Fenvalerate and their structural analogs bifenthrin, deltamethrin, homeopathic permethrin and four other pesticides commonly used in agricultural production were selected for cross-reaction experiments. Ten pesticides were configured with the same concentration as the standard and competed with the capture probe for the signal probe. There were detected and recorded at a fluorescence value of 542 nm. The fluorescence intensity after the reaction of the capture probe and the signal probe was recorded as I_0, the fluorescence intensity after the reaction of different standards with the capture probe and the signal probe was recorded as I, and the difference between I_0 and I was calculated as ΔI.

3. Results and Discussion

3.1. Characterization of Upconversion Materials

Rare earth doped upconversion nanomaterials have excellent physical and chemical properties, so they have been widely used. However, the low luminous efficiency had become the biggest resistance to the research and development of upconversion nanoparticles. The design of core–shell structure could greatly make up for the defects of upconversion materials. Rare earth upconversion materials are generally composed of matrix, sensitizer and activator. The choice of matrix materials could be oxides, sulfides, halides, etc., but the high phonon energy of oxides was not conducive to the occurrence of upconversion, and the iodide and bromide in halides were easy to be decomposed by moisture, which was not conducive to the preparation of materials. Therefore, fluoride was selected as the matrix material. $NaYF_4$ in fluoride had good stability and high emission efficiency and was the first choice of matrix materials. The rare earth ions doped with upconversion nanomaterials had low photon utilization in the near-infrared region. If the absorption efficiency was increased by increasing the concentration of doped ions, the distance between activator ions was too close, and cross-relaxation will occur, resulting in concentration quenching and the luminous efficiency could not be improved. To improve the luminescence efficiency, a higher proportion of sensitizer ions could be added to the upconversion nanocrystals. Compared with other sensitizer ions, Yb^{3+} had the widest absorption cross-section in the near-infrared region. The activator ion Er^{3+} not only had a rich energy level structure but also had a narrow luminescence band, which could have efficient energy transfer with the sensitizer ion Yb^{3+}. Due to the surface of rare earth upconversion nanocrystalline $NaYF_4$:Yb,Er@$NaYF_4$ being in direct contact with solvent and environment, a large number of ions were able to reach the exciting state transfer energy to the surface of upconversion nanoparticles and return to the ground state in a nonradiative transition, resulting in surface quenching, which greatly affects the luminescence efficiency of upconversion. Therefore, to suppress the occurrence of surface quenching, a layer of inert material was usually coated on the outside. As an inert shell, $NaYF_4$ could not only have better lattice matching with the template seed crystal and better coating effect, but also because $NaYF_4$ is an inert material, it inhibits the nonradiative relaxation of excited Er^{3+}, making the energy transfer between Er^{3+} for a longer time, and increased the fluorescence lifetime of the material. Therefore, inert core/active shell/inert core three-layer sandwich structure upconversion nanoparticles $NaYF_4$@$NaYF_4$:Yb, Er@$NaYF_4$ were designed.

The morphology and size of upconversion nanomaterials were measured by transmission electron microscope. As shown in Figure 1, upconversion particles had good dispersion, uniform size and hexagonal phase. Figure 1A showed the inert core $NaYF_4$ with a particle size of about 16 nm, Figure 1B showed the morphology of the inert core $NaYF_4$ coated with the active shell, $NaYF_4$:Yb, Er@$NaYF_4$ about 22 nm, Figure 1C showed the morphology of $NaYF_4$: Yb, Er@$NaYF_4$ coated with the inert shell, and $NaYF_4$@$NaYF_4$:Yb, Er@$NaYF_4$ about 26 nm. The coating of the core–shell structure could be proved to be successful by increasing the particle size. The fluorescence spectrum of UCNPs was measured by using a fluorescence spectrophotometer with an external 980 nm excitation light source. It could be seen from Figure 1E that the nanoparticles had the largest characteristic emission peak at 542 nm. The fluorescence efficiency of the nanomaterial was significantly enhanced after coating the active shell, and the fluorescence intensity increased several times after coating the inert shell. Compared with the core–shell structure $NaYF_4$:Yb,Er@$NaYF_4$ synthesized by Nahid ghazyani [23], the fluorescence intensity increased nearly 100 times. The coating of the inert shell greatly shortened the distance between the luminous center and the energy receptor, shortened the path from the sensitizer to the activator, reduced the energy loss, had high quantum efficiency, and realized high-strength emission.

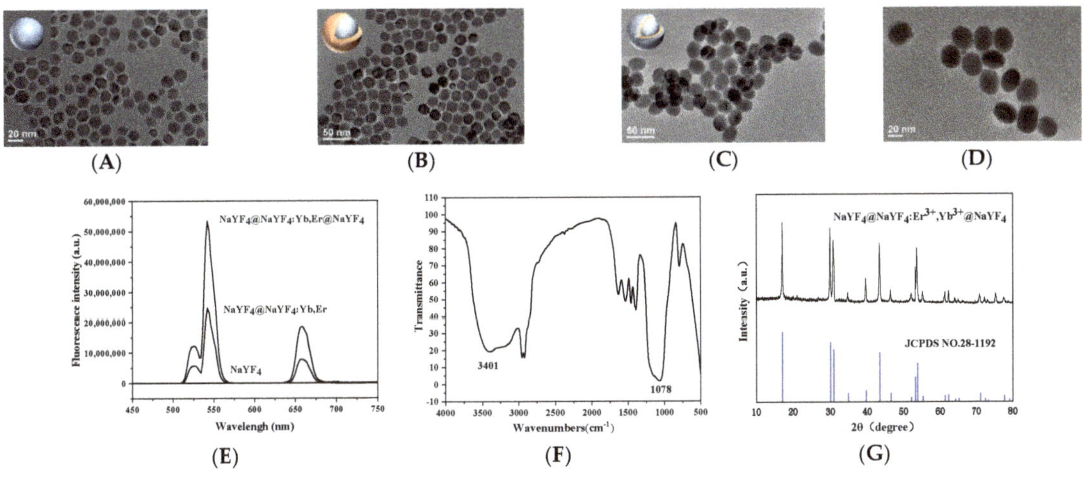

Figure 1. (**A**) TEM images of $NaYF_4$; (**B**) TEM images of $NaYF_4$:Yb,Er@$NaYF_4$; (**C**) TEM images of $NaYF_4$@$NaYF_4$:Yb,Er@$NaYF_4$; (**D**) TEM images of $NaYF_4$@$NaYF_4$:Yb,Er@$NaYF_4$@SiO_2@NH_2; (**E**) Flu-orescence spectra; (**F**) FTIR spectra of $NaYF_4$@$NaYF_4$:Yb,Er@$NaYF_4$@SiO_2@NH_2; and (**G**) XRD pat-tern of $NaYF_4$@$NaYF_4$:Yb,Er@$NaYF_4$.

To increase the dispersion and biocompatibility of upconversion materials in the aqueous system, they were modified with a layer of SiO_2 shell and amino group on the surface by the classical Stober method. Figure 1D shows the transmission electron microscope image of the modified upconversion material. It could be seen from the figure that a thin layer of silicon shell appears on the surface of the modified UCNPs, and the modified particle size was about 30 nm, which indicated that the silicon shell on the surface of UCNPs was coated successfully. The functional groups on the surface of the modified nanomaterial were characterized by Fourier infrared spectroscopy. As shown in Figure 1F, the siloxy symmetric stretching and stretching vibration of the nanoparticle appeared in the area of about 1100 cm^{-1}, which again showed that the material had been coated with a silicon shell. The stretching and bending vibration of the amino group appeared at 3401 cm^{-1}, indicating that the surface of the material had been modified with the amino group. The peak at 2900 cm^{-1} in the figure was methylene symmetrical stretching vibration, which may be due to accidental contamination with substances containing methylene during tablet pressing, which will not affect the experimental results. The results of X-ray diffraction of the nanomaterial are shown in Figure 1G, and the diffraction peak was consistent with the comparison result of hexagonal beta phase UCNPs standard spectrum (JCPDS No. 28-1192), which further showed that the synthesized UCNPs was a hexagonal phase. The phonon energy of the hexagonal phase was lower than that of the cubic phase, which was helpful to improve the upconversion luminous efficiency of the material and was more suitable to be used as an upconversion matrix material.

3.2. Optimization of Antigen and Antibody Addition

Based on the principle of immunology, the signal probe was made by connecting pyrethroid universal antibody with modified upconversion particles by glutaraldehyde cross-linking method, and the capture probe was made by connecting pyrethroid universal antigen with amino magnetic microspheres.

Figure 2 shows the detection results of antigen–antibody concentration in the supernatant after universal antigen and antibody were combined with magnetic particles and upconversion materials respectively. As shown in Figure 2B, when the addition amount of pyrethroid universal antigen was 20~80 μg, the coupling rate increases with the increase of the additional amount. When the addition amount was 80 μg, the coupling rate between antigen and amino magnetic microspheres reached the maximum. When the addition amount was 80~100 μg, the coupling rate decreased gradually. So when 80 μg antigen was added, the antigen combined with the amino magnetic microspheres performed best, and the coupling rate was 97.01%. Figure 2C shows that when the addition amount of pyrethroid universal antibody was 20~80 μg, the coupling rate increased gradually. When the amount of antibody added was 80 μg, the coupling rate between antibody and upconversion material reached the maximum. When the addition amount was 80~100 μg, the coupling rate decreased gradually. Therefore, 80 μg antibody was added when preparing the signal probe, and the coupling rate was 95.06%.

3.3. Optimization of Capture Probe

In the upconversion magnetic separation detection system, the capture probe was used as the antigen carrier and separation medium. If the addition amount was insufficient, it will affect the full binding of antigen and antibody, and finally lead to the weak fluorescence signal, so the fluorescence spectrophotometer could not detect the change of fluorescence signal difference. If the addition was excessive, the color of the reaction system will be darker, which may cause background interference, affect the fluorescence value and lead to waste.

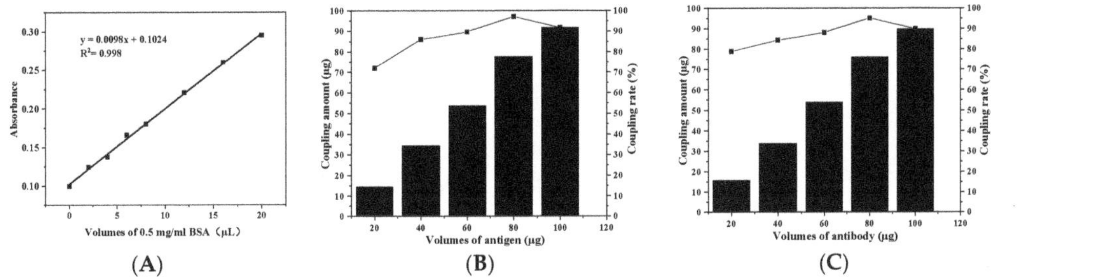

Figure 2. (**A**) BCA kit standard curve; (**B**) Optimization of the added amount of pyrethroid universal antigen conjugated with MPMs; and (**C**) Optimization of the added amount of pyrethroid universal antibody conjugated with UCNPs.

On the premise of determining the additional amount of signal probe, the addition amount of capture probe was optimized. The optimization results of the addition amount of capture probe are shown in Figure 3. When the addition amount of capture probe was 40~120 μg, the fluorescence value of the capture probe signal probe complex gradually increased. When the addition amount of the capture probe was 120 μg, the fluorescence value of the reaction system reached the maximum. When the addition amount of capture probe was 120~140 μg, the fluorescence value of the capture probe signal probe complex decreased gradually. To achieve the best detection results without wasting materials, the optimal addition amount of capture probe was 120 μg.

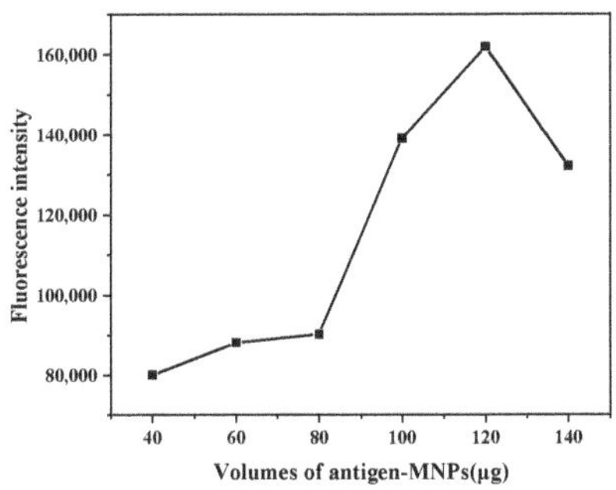

Figure 3. Optimization of capture probe addition.

3.4. *Drawing of Standard Curve and Determination of Detection Limit*

For this procedure, 50 μL standard solutions of fenpropathrin, cypermethrin and fenvalerate of different concentrations and 120 μL capture probe were prepared, placed into the centrifuge tubes, to which 200 μL signal probe was added, and subjected to competitive reaction for 60 min. The signal probe capture probe conjugate was collected by the magnet and washed 3 times with PBS. The precipitate was dispersed in 2 mL PBS again, and the fluorescence intensity at 542 nm was measured by a fluorescence spectrophotometer with an external 980 nm laser. The logarithm was taken of the concentrations of fenpropathrin, cypermethrin and fenvalerate as the abscissa and the fluorescence intensity difference ($\Delta I = I_0 - I$) of immune complexes in the detection system containing different concentrations of standards as the ordinate, and the standard curve for the detection of fenpropathrin, cypermethrin and fenvalerate was established. The fluorescence spectra at different concentrations of fenpropathrin, cypermethrin and fenvalerate are shown in Figure 4. With the increase of standard concentration, the fluorescence intensity in the reaction system decreased. According to the standard curve, it could be found that the LOD and sensitivity of upconversion magnetic separation immunoassay for fenpropathrin were 0.01 μg/L and 1.6 μg/L, for cypermethrin were 0.015 μg/L and 1.3 μg/L, and for fenvalerate were 0.011 μg/L and 1.5 μg/L. Liu Mengli [24] constructed a label-free fluorescence analysis method using $NaYF_4$:Yb,Er. The LOD of fenpropathrin in traditional Chinese medicine was 0.24 ng/mL, the fluorescence intensity of $NaYF_4$@$NaYF_4$:Yb,Er@$NaYF_4$ synthesized by this method was higher, and the LOD of upconversion magnetic separation immunoassay for pyrethroids was lower. This immunoassay method was more convenient, sensitive and efficient than most traditional methods for the detection of pyrethroids. The research of this method provides a powerful means for the detection of pyrethroids in food.

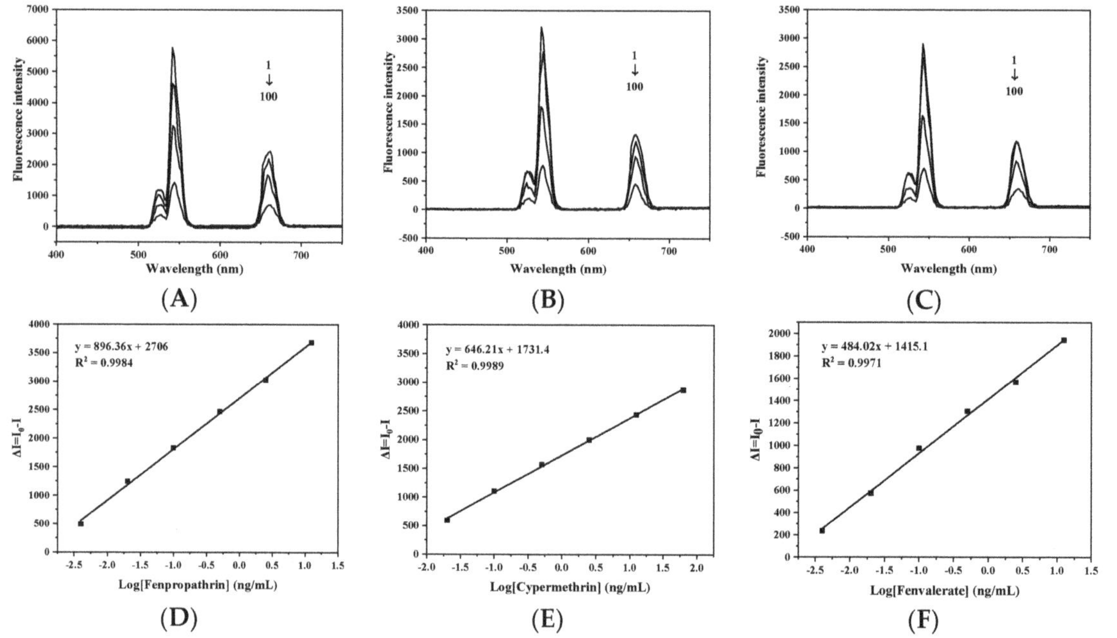

Figure 4. (**A**) Upconversion fluorescence spectrum of fenpropathrin at different concentrations; (**B**) Upconversion fluorescence spectrum of cypermethrin at different concentrations; (**C**) Upconversion fluorescence spectrum of fenvalerate at different concentrations; (**D**) Standard curve of fenpropathrin; (**E**) Standard curve of cypermethrin; and (**F**) Standard curve of fenvalerate.

3.5. Elimination of Matrix Effect

Proteins, sugars and pigments in the sample will interfere with the test results. The influence of the matrix could be reduced by sample dilution. Apple, pear, cabbage and cucumber samples were extracted with acetonitrile and diluted with PBS 5, 10 and 20 times respectively. The standard inhibition curve was drawn with the concentration of pyrethroids as the abscissa and the fluorescence intensity as the ordinate. The results are shown in Figure 5. When cucumber and cabbage samples were diluted 10 times and apple and pear were diluted 20 times, the standard inhibition curve was similar to the standard curve of the established method. It showed that the matrix effect could be eliminated when the cabbage and cucumber were diluted 10 times, and apple and pear were diluted 20 times.

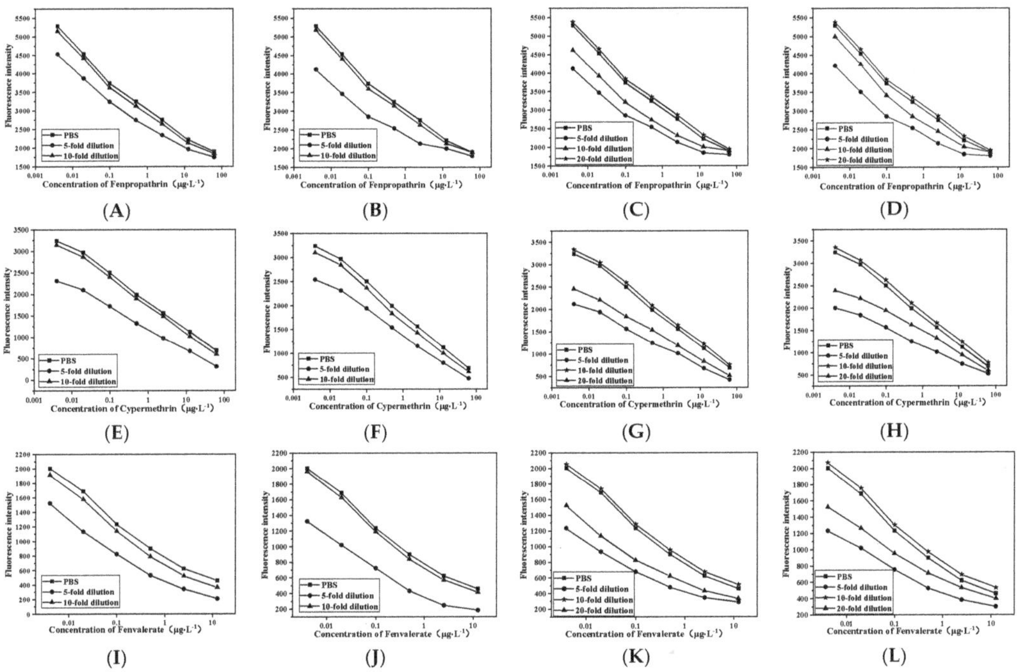

Figure 5. (**A**,**E**,**I**) Elimination results of matrix effect of cucumber; (**B**,**F**,**J**) cabbage; (**C**,**G**,**K**) apple; and (**D**,**H**,**L**) pear.

3.6. Add Sample Recovery

All samples were detected by HPLC without pyrethroids before the addition and recovery experiment. Pyrethroid standards with low, medium and high concentrations were added to cucumber, cabbage, apple and pear respectively, and five parallel groups were set for the determination of each concentration of each sample. The same amount of standard with the same concentration was added to the same five samples, and the concentration of standard was determined by the method established in this paper. After sample pretreatment and PBS dilution, they were used for sample determination. Calculation of the recovery and CV of spiked samples was made according to the formula. As shown in Table 1, the average recovery of fenpropathrin was 88.6~97.8% and CV was 2.1~7.1%; the average recovery of cypermethrin was 84.5~93.8% and CV was 2.1~7.4%; the average recovery of fenvalerate was 83.4~94.2% and CV was 1.9~6.8%. The CV was not higher than 7.4%, indicating that the method established in this paper had good precision. The detection limit of apple and pear samples was 1 $\mu g \cdot kg^{-1}$, the detection limit of cabbage and cucumber was 0.5 $\mu g \cdot kg^{-1}$; it was far lower than the national standard of MRL in fruits and vegetables [21]. It showed that the method has good accuracy and could be used for the detection of pyrethroids in daily life.

$$\text{rate of recovery\%} = \frac{\text{Detection amount of spiked sample}}{\text{Standard addition}} \times 100\% \tag{2}$$

$$CV = \frac{SD}{MN} \times 100\% \tag{3}$$

Table 1. Recovery rate of pyrethroids in added samples.

Sample	Additive Concentration (μg·kg^{-1})	Fenpropathrin Mean ± SD (μg·kg^{-1})	Fenpropathrin Average Recovery (%)	Fenpropathrin CV (%)	Cypermethrin Mean ± SD (μg·kg^{-1})	Cypermethrin Average Recovery (%)	Cypermethrin CV (%)	Fenvalerate Mean ± SD (μg·kg^{-1})	Fenvalerate Average Recovery (%)	Fenvalerate CV (%)
Cucumber	0.5	0.44 ± 0.03	88.6	7.1	0.42 ± 0.03	84.5	7.4	0.42 ± 0.03	83.4	6.7
	2	1.83 ± 0.08	91.5	4.6	1.72 ± 0.07	86.2	4.1	1.71 ± 0.06	85.4	3.6
	10	9.31 ± 0.27	93.1	2.9	8.87 ± 0.19	88.7	2.1	8.79 ± 017	87.9	1.9
Cabbage	0.5	0.46 ± 0.03	92.7	6.1	0.45 ± 0.03	89.4	5.7	0.46 ± 0.03	92.6	5.8
	2	1.93 ± 0.09	96.5	4.5	1.83 ± 0.08	91.5	4.1	1.87 ± 0.07	93.4	3.9
	10	9.78 ± 0.36	97.8	3.7	9.33 ± 0.27	93.3	2.9	9.42 ± 0.23	94.2	2.4
Apple	0.5	0.44 ± 0.03	88.7	7.1	0.45 ± 0.03	89.6	6.9	0.45 ± 0.03	89.5	6.8
	2	1.83 ± 0.08	91.6	4.6	1.84 ± 0.07	92.1	3.9	1.81 ± 0.07	90.7	3.6
	10	9.51 ± 0.27	95.1	2.8	9.38 ± 0.27	93.8	2.9	9.29 ± 0.23	92.9	2.5
Pear	0.5	0.45 ± 0.03	89.7	5.9	0.44 ± 0.03	87.8	5.7	0.44 ± 0.03	88.7	5.8
	2	1.84 ± 0.08	91.9	4.2	1.78 ± 0.07	89.2	3.9	1.80 ± 0.07	89.9	4.1
	10	9.43 ± 0.20	94.3	2.1	9.18 ± 0.26	91.8	2.8	9.23 ± 0.22	92.3	2.4

In the formula, CV is the coefficient of variation, SD is the standard deviation, and MN is the average value.

3.7. Comparison Experiment of High-Performance Liquid Chromatography

To verify the accuracy of upconversion magnetic separation immunoassay, the concentration of pyrethroids in blind samples was determined by HPLC. One sample was divided into two parts, and the same amount of standard was added respectively. One part was pretreated according to the description in part 2.6, and the method established in this paper was used for detection; another part of the samples was pretreated as described in Section 2.7 and detected by HPLC. Observation of the correlation between the results of the two detection methods was made. The retention time of blank spiked samples was 16.5 min for fenpropathrin, 18.7 min for cypermethrin and 21.3 min for fenvalerate. The correlation between the HPLC method and this research method is shown in Figure 6; the R^2 of the correlation equation was greater than 0.995, indicating that the two methods had a good consistency. It showed that the method established in this paper is accurate and reliable. The upconversion magnetic separation immunofluorescence assay was accurate, and could be used for the detection of pyrethroids in real samples.

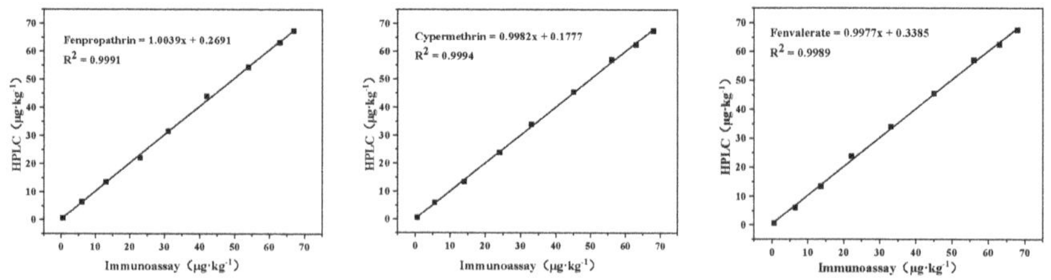

(A) (B) (C)

Figure 6. Correlation between HPLC and this research method in the detection of (**A**) fenpropathrin; (**B**) cypermethrin and (**C**) fenvalerate.

3.8. Actual Sample Test Results

The prepared blind samples were tested by this experimental method. The results are shown in Table 2. Three positive samples were detected in three cucumber blind samples; among the three blind samples of cabbage, two positive samples were detected, and the rest were negative samples; two positive samples were detected in three blind apple samples, and the rest were negative samples; three positive samples were detected in three blind pear samples. The results obtained by HPLC and this experimental method were compared, and the number of positive samples detected by the two methods was the same. The accuracy and practical applicability of the method established in this paper were evaluated by adding the recovery experiment and the actual sample detection experiment. The results showed that the upconversion magnetic separation immunofluorescence assay established in this experiment had good accuracy and could be applied to the detection of pyrethroid pesticide residues in vegetables.

Table 2. Detection of pyrethroids in actual samples.

	This Experimental Method			HPLC		
Sample	Fenpropathrin ($mg \cdot kg^{-1}$)	Cypermethrin ($mg \cdot kg^{-1}$)	Fenvalerate ($mg \cdot kg^{-1}$)	Fenpropathrn ($mg \cdot kg^{-1}$)	Cypermethrin ($mg \cdot kg^{-1}$)	Fenvalerate ($mg \cdot kg^{-1}$)
1#cucumber	-	-	85.6	-	-	86.4
2#cucumber	-	65.8	-	-	67.1	-
3#cucumber	57.3	-	-	58.6	-	-
4#cabbage	-	-	-	-	-	-
5#cabbage	78.2	-	-	79.5	-	-
6#cabbage	-	-	86.1	-	-	87.3
7#apple	-	-	46.1	-	-	47.6
8#apple	-	85.6	-	-	86.4	-
9#apple	-	-	-	-	-	-
10#pear	-	45.7	-	-	46.5	-
11#pear	-	-	66.5	-	-	67.9
12#pear	85.4	-	-	86.7	-	-

"-" indicates negative sample.

3.9. Method Specificity Evaluation

As shown in Figure 7, when the added standard was fenpropathrin, cypermethrin and fenvalerate, the change range of fluorescence intensity was large, indicating that the signal probe was specifically combined with fenpropathrin, cypermethrin and fenvalerate. When the added standard was its structural analogs bifenthrin, deltamethrin and cis-permethrin, the change range of fluorescence intensity was relatively large, indicating that the signal probe can also be combined with other pyrethroid standards, this method was suitable for the detection of pyrethroids. When the standard was other pesticides, the change range of fluorescence intensity was small, the signal probe had almost no specific binding with the standard, and there was only some nonspecific adsorption.

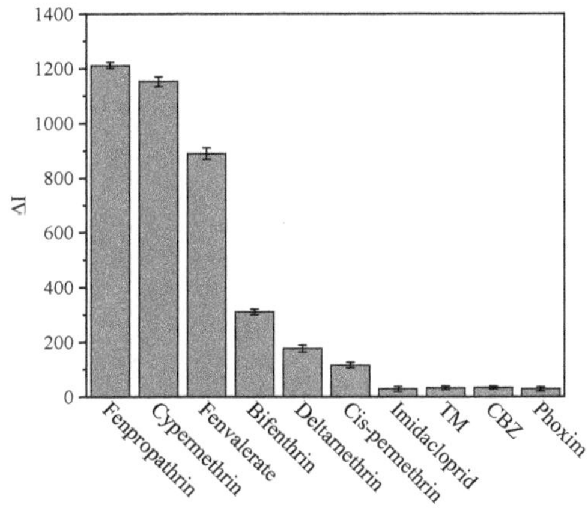

Figure 7. Cross reaction.

4. Conclusions

In this study, a new three-layer core-shell structure upconversion nanoparticle was synthesized. The activated material was coupled with pyrethroid antibody to prepare a signal probe, and the amino polystyrene magnetic microspheres were coupled with pyrethroid antigen to prepare a capture probe. Based on the specific binding of antigen and antibody and the characteristics that the magnetic microspheres could be separated by the external magnetic field, an upconversion magnetic separation immunofluorescence method for detecting pyrethroid residues was established. The upconversion magnetic separation immunofluorescence method for the detection of pyrethroid residues established in this paper had higher accuracy and lower detection limit than enzyme-linked immunosorbent assay; compared with the large-scale instrument detection method, this method has the advantages of being fast, convenient and low cost. The establishment of the upconversion magnetic separation immunofluorescence method enriches the technical means of the rapid detection market and provides theoretical support for the detection of pyrethroid residues.

Author Contributions: Conceptualization, Y.S.; methodology, L.Z.; software, W.Z.; validation, W.Z.; formal analysis, J.J.; investigation, L.Z.; resources, Y.S.; data curation, J.J.; writing—original draft preparation, L.Z.; writing—review and editing, Y.S. and Y.Z.; visualization, J.J.; supervision, Y.S.; project administration, L.Z.; funding acquisition, Y.S. All authors have read and agreed to the published version of the manuscript.

Funding: This research was funded by Open Project Program of State Key Laboratory of Food Nutrition and Safety, Tianjin University of Science and Technology (No. SKLFNS-KF-391 202115) and the Natural Science Foundation of Tianjin (18JCQNJC84400) and National Natural Science Foundation (No. 31301487).

Institutional Review Board Statement: Not applicable.

Informed Consent Statement: Not applicable.

Data Availability Statement: Not applicable.

Acknowledgments: Thanks to Mingfei Pan for his help in this research.

Conflicts of Interest: The authors declare no conflict of interest.

References

1. Qin, Y.K.; Li, S.; Hong, M.; Wang, Z.G.; Ji, G.G.; Chen, R.P.; Zhao, X.D.; Wang, Y.; Ren, S.Y.; Han, D.P.; et al. Synthesis and functionalization of upconversion nano materials and their application in food safety detection. *Anal. Chem.* **2021**, *49*, 1955–1969.
2. Shi, F.; Wang, X.; Liu, M. β-NaYF_ 4: Preparation and luminescent properties of Yb, X% ho materials. *Guangdong Chem. Ind.* **2021**, *48*, 28–29+37.
3. Hong, A.R.; Kim, Y.; Lee, T.S.; Kim, S.; Lee, K.; Kim, G.; Jang, H.S. Intense Red-Emitting Upconversion Nanophosphors (800 nm-Driven) with a Core/Double-Shell Structure for Dual-Modal Upconversion Luminescence and Magnetic Resonance in Vivo Imaging Applications. *Acs Appl. Mater. Interfaces* **2018**, *10*, 12331–12340. [CrossRef] [PubMed]
4. Li, X.X.; Liu, L.; Fu, Y.; Chen, H.D. Peptide-enhanced tumor accumulation of upconversion nanoparticles for sensitive upconversion luminescence/magnetic resonance dual-mode bioimaging of colorectal tumors. *Acta Biomater.* **2020**, *104*, 167–175. [CrossRef] [PubMed]
5. Zhang, W.N.; Li, J.; Lei, H.X.; Li, B.J. Temperature-dependent Forster resonance energy transfer from upconversion nanoparticles to quantum dots. *Opt. Express* **2020**, *28*, 12450–12459. [CrossRef]
6. Feng, A.L.; Xu, R.; Wang, Y.N.; Zhang, Y.N.; Lin, S.B. Core shell rare earth upconversion nano materials and their biomedical applications. *Mater. Guide* **2019**, *33*, 2252–2259.
7. Wu, H.Y. Yb ~ (3+)/Er ~ (3+) Co Doped NaYF_ 4 Nanocrystalline Core-Shell Structure and Its Fluorescence Characteristics. Master's Thesis, Harbin Institute of Technology, Harbin, China, 2020.
8. Boyer, J.C.; Veggel, F.V. Absolute quantum yield measurements of colloidal NaYF4: Er3+, Yb3+ upconverting nanoparticles. *Nanoscale* **2010**, *2*, 1417–1419. [CrossRef]
9. Zhu, Y.Y. Preparation and Fluorescence Properties of NaYbF_4@NaYF_4: Yb/Er. Master's Thesis, Harbin Institute of Technology, Harbin, China, 2020.
10. Qiu, H.L.; Yang, C.H.; Shao, W.; Damasco, J.; Wang, X.L.; Agren, H.; Prasad, P.N.; Chen, G.Y. Enhanced Upconversion Luminescence in Yb^{3+}/Tm^{3+}-Codoped Fluoride Active Core/Active Shell/Inert Shell Nanoparticles through Directed Energy Migration. *Nanomaterials* **2014**, *4*, 55–68. [CrossRef] [PubMed]

11. Ji, X.Y.; Ye, T.; Yuan, M.; Cao, H.; Xu, W.; Yu, J.S.; Ren, H.; Ye, Y.Y.; Li, J.Y.; Wang, Y.X. Preparation and application of molecularly imprinted fluorescent sensor for pyrethroid pesticide metabolites. *J. Anal. Test.* **2019**, *38*, 1487–1492.
12. Peng, Y.Q.; Li, Z.H.; Liu, Z.I.; Yuan, Q. Biomedical applications of core-shell rare earth upconversion nano materials. *Chin. Sci. Chem.* **2015**, *45*, 1159–1177.
13. Richardson, J.R.; Fitsanakis, V.; Westerink, R.H.S.; Kanthasamy, A.G. Neurotoxicity of pesticides. *Acta Neuropathol.* **2019**, *138*, 343–362. [CrossRef] [PubMed]
14. Al-Ghanim, K.A.; Mahboob, S.; Vijayaraghavan, P.; Al-Misned, F.A.; Kim, Y.O.; Kim, H.J. Sub-lethal effect of synthetic pyrethroid pesticide on metabolic enzymes and protein profile of non-target Zebra fish, Danio rerio. *Saudi J. Biol. Sci.* **2020**, *27*, 441–447. [CrossRef] [PubMed]
15. Bravo, N.; Grimalt, J.O.; Bocca, B.; Pino, A.; Bin, M.; Brumatti, L.V.; Rosolen, V.; Barbone, F.; Ronfani, L.; Alimonti, A.; et al. Urinary metabolites of organophosphate and pyrethroid pesticides in children from an Italian cohort. *Environ. Res.* **2019**, *32*, 108–117. [CrossRef] [PubMed]
16. Curl, C.L.; Porter, J.; Penwell, I.; Phinney, R.; Ospina, M.; Calafat, A.M. Effect of a 24-week randomized trial of an organic produce intervention on pyrethroid and organophosphate pesticide exposure among pregnant women. *Environ. Int.* **2019**, *115*, 104–113. [CrossRef] [PubMed]
17. Pasupuleti, R.R.; Tsai, P.C.; Ponnusamy, V.K. A fast and sensitive analytical procedure for monitoring of synthetic pyrethroid pesticides' metabolites in environmental water samples. *Microchem. J.* **2019**, *148*, 355–363. [CrossRef]
18. Lu, Q.; Li, H.D.; Ding, R.Y.; Guo, C.Y.; Fang, L.P.; Zhang, W.J.; Mao, J.S.; Chen, Z.L. Determination of pyrethroid residues in Pear by quers-gc. *Pesticides* **2019**, *58*, 356–358.
19. Zhao, H.M.; Bian, H.Z.; Fang, L. Determination of bifenthrin and deltamethrin pesticide residues in vegetables and fruits by quechers-gc-ms/MS. *Anhui Agron. Bull.* **2019**, *25*, 94–97.
20. Chrustek, A.; Holynska-Iwan, I.; Dziembowska, I.; Bogusiewicz, J.; Wroblewski, M.; Cwynar, A.; Olszewska-Slonina, D. Current Research on the Safety of Pyrethroids Used as Insecticides. *Med.-Lith.* **2018**, *54*, 61. [CrossRef] [PubMed]
21. *GB 2763-2021*; National Food Safety Standard—Maximum Residue Limits for Pesticides in Food. Available online: http://down.foodmate.net/wap/index.php?itemid=97819&moduleid=23(accessed on 9 March 2022).
22. *GB 23200.8-2016*; Method for Determination of 500 Pesticides and Related Chemicals Residues in Fruits and Vegetables-GC-MS Method. Available online: https://www.doc88.com/p-0874842560376.html(accessed on 9 March 2022).
23. Ghazyani, N.; Ara, M.H.M.; Raoufi, M. Nonlinear photoresponse of NaYF4:Yb,Er@NaYF4 nanocrystals under green CW excitation: A comprehensive study. *RSC Adv.* **2020**, *10*, 25696–25702. [CrossRef]
24. Liu, M.L. Study on Fluorescence up Conversion Nano Materials for the Detection of Pesticide Residues in Traditional Chinese Medicine and Their Analytical Application. Master's Thesis, Southwest University, Chongqing, China, 2019.

foods

Article

Preparation of Core-Shell Rare Earth-Doped Upconversion Nanomaterials and Simultaneous Detection of Two Pesticides in Food

Wenbo Zhu [1,†], Lingyan Zhao [1,†], Jingyi Jin [1] and Yang Song [1,2,*]

1 Tianjin Key Laboratory of Animal and Plant Resistance, College of Life Sciences, Tianjin Normal University, Tianjin 300387, China; 1910170006@stu.tjnu.edu.cn (W.Z.); 2010170003@stu.tjnu.edu.cn (L.Z.); 2110170005@stu.tjnu.edu.cn (J.J.)
2 State Key Laboratory of Food Nutrition and Safety, College of Food Science and Engineering, Tianjin University of Science and Technology, Tianjin 300457, China
* Correspondence: skysongy@mail.tjnu.edu.cn; Tel.: +86-22-2376-5561
† These authors contributed equally to this work.

Abstract: Under the excitation of a 980 nm excitation light, the fluorescence signals of the synthesized core-shell $NaYF_4$:Yb@$NaYF_4$:Ho and monolayer $NaYF_4$:Yb,Tm upconversion nanoparticles (UCNPs) were simultaneously detected at 656 and 696 nm, respectively. The two upconversion materials were coupled with anti-clothianidin and anti-imidacloprid monoclonal antibodies by the glutaraldehyde cross-linking method as signal probes. Imidacloprid (IMI) and clothianidin (CLO) could compete with antigen-conjugated amino Fe_3O_4 magnetic nanomaterials for binding to signaling probes, thus establishing a rapid and sensitive fluorescent immunoassay for the simultaneous detection of IMI and CLO. Under optimal conditions, the limits of detection (LOD, IC_{10}) and sensitivity (IC_{50}) of IMI and CLO were (0.032, 0.028) and (4.7, 2.1) ng/mL, respectively, and the linear assay ranges were at 0.032–285.75 ng/mL and 0.028–200 ng/mL, respectively. Immunoassay did not cross-react significantly with other analogs. In fruits and vegetables such as apples, oranges, peaches, cucumbers, tomatoes and peppers, the mean recoveries of IMI and CLO ranged from 83.33% to 115.02% with relative standard deviations (RSDs) of 1.9% to 9.2% and 1.2% to 9.0%, respectively. Furthermore, the results of the immunoassay correlate well with the high-performance liquid chromatography method used to detect the actual samples.

Keywords: upconversion nanomaterials; core-shell structure upconversion nanomaterials; imidacloprid; clothianidin; food

Citation: Zhu, W.; Zhao, L.; Jin, J.; Song, Y. Preparation of Core-Shell Rare Earth-Doped Upconversion Nanomaterials and Simultaneous Detection of Two Pesticides in Food. *Foods* **2022**, *11*, 1485. https://doi.org/10.3390/foods11101485

Academic Editor: Alberto Angioni

Received: 2 April 2022
Accepted: 17 May 2022
Published: 19 May 2022

1. Introduction

Neonicotinoids (NEOs) are the fourth largest class of pesticides after organophosphates, carbamates and pyrethroids. They have the characteristics of high compatibility with the environment and no cross-resistance with other traditional pesticides [1]. As of 2012, more than 120 countries have registered the use of neonicotinoid pesticides. IMI (36.9%) and CLO (14.3%) account for more than half of the total share of these pesticides [2]. However, such pesticides were constantly detected in food, soil and groundwater, which brought great potential safety hazards to human health and the ecological environment [3,4] and had a certain negative impact on pollinators such as bees [5,6]. They have also had lethal effects on aquatic and terrestrial invertebrates [7,8]. In addition, IMI and CLO have also had potential toxic risks such as hepatotoxicity, neurotoxicity and genotoxicity, which have attracted wide attention from countries around the world [9,10].

In response to the problems of the above neonicotinoid pesticides, China's GB 2763-2021 "National Food Safety Standard Food Maximum Residue Limits for Pesticides" [11] stipulated that the maximum residue limits (MRLs) of IMI and CLO in fruits and vegetables should be 0.05–5 mg/kg and 0.01–2 mg/kg. The EU stipulated that the MRLs of

IMI [12] and CLO [13] should be 0.05–5 mg/kg and 0.01–1.5 mg/kg in fruits and vegetables, respectively. In addition, the United States [14], Canada [15] and other countries have also banned and restricted such pesticides. The current detection techniques for neonicotinoid pesticides include liquid chromatography [16], liquid chromatography-tandem mass spectrometry [17], and enzyme-linked immunosorbent assay [18]. Li Dongpo et al. [19] established a high-performance liquid chromatography-tandem mass spectrometry method for the determination of IMI and emamectin-benzoate residue in honeysuckle, with quantitative limits of 2.5 and 10 μg/kg, respectively, which provides a theoretical basis for the detection of such pesticides in traditional Chinese medicine. But its pretreatment is relatively cumbersome and time-consuming. Watanabe et al. [18] used the enzyme-linked immunosorbent assay (ELISA) to detect dinotefuran, IMI and CLO in leeks. The average recovery of spiked samples was 85–113%, and the relative standard deviation was less than 15%. However, ELISA has the disadvantages of many operation steps and false-positive results. Hua et al. [20] established a fluorescence immunoassay for the detection of IMI by using the synthesized $NaYF_4$/Yb,Er UCNPs. The IC_{50} and detection limit (IC_{10}) of IMI were 14.59 and 0.74, respectively. It can only detect a single target, which has certain limitations. Instrumental method detection requires professional operation and consumes a lot of solvents. The enzyme-linked immunosorbent assay requires multi-step incubation and washing procedures, which are cumbersome and time-consuming, and cannot meet the rapid monitoring of the market. Therefore, it is of great practical significance to establish a sensitive, efficient and convenient method for the detection of new nicotine pesticides.

Upconversion nanoparticles (UCNPs) could emit strong visible fluorescence under the excitation of a 980 nm light source and had the advantages of low toxicity, good chemical stability, a narrow emission band gap, and low fluorescence background interference [21,22]. However, there was still the problem that the surface is prone to fluorescence quenching. Core-shell structured upconversion nanomaterials formed by coating the core with one or more layers of materials may solve the problem [23] since the shell deposition may reduce the quenching effect caused by surface defects such as the suspension key and vibration groups and thus, effectively improve the luminous efficiency [21]. In addition, by rationally assigning the spatial position and relative concentration of the activator, precise upconversion luminescence color adjustment could also be achieved [24,25].

At present, $NaYF_4$:Yb,Er, $NaYF_4$:Yb,Tm and other UCNPs based on $NaYF_4$ are the research focus in the field of nanomaterials. Excited by 980 nm light, $NaYF_4$:Yb,Er and $NaYF_4$:Yb,Tm emit green light and blue light at 540 nm and 470 nm, respectively, but the penetration depth of these two lights is small [26]. The light wavelength applied to deep-seated imaging of biological tissues is usually more than 600 nm, and the corresponding emitted light is red. At present, the UCNPs of Yb/Er and Yb/Tm co-doped systems are the main research objects of red-emitting materials, and there is little research on Yb/Ho system. Therefore, the study of Yb/ Ho co-doped UCNPs emitting red light can expand the application of biomarkers and biological imaging, which has important scientific significance and practical value. Research on core-shell structure upconversion materials is still in its infancy, and although we found many research papers in this area, there were few papers applying it to immunoassay detection. In this study, a new core-shell nanomaterial $NaYF_4$:Yb@$NaYF_4$:Ho was introduced into immunoassay for the first time, and a fluorescence immunoassay method based on magnetic separation and upconversion nanoparticles was established. This method has high sensitivity and accuracy and can be used for the simultaneous determination of IMI and CLO in food.

2. Materials and Methods

2.1. Materials and Reagents

IMI, CLO, acetamiprid (ACE), thiacloprid (THI), nitenpyram (NIT), dinotefuran (DIN), thiamethoxam (THX), ytterbium chloride, yttrium chloride, thulium chloride, holmium chloride, ammonium fluoride, oleic acid, cyclohexane, 3-aminopropyltriethoxysilane (APTES, 98%), tetraethyl orthosilicate (TEOS, 98%) were purchased from Adamas (Shanghai, China).

Argon was purchased from Air Liquide Co., Ltd. (Tianjin, China). Bovine serum albumin (BSA, 98.0%) was purchased from Merck, Darmstadt, Germany. IMI antigen, CLO antigen, anti-IMI monoclonal antibody and anti-CLO monoclonal antibody were purchased from Shandong Landu Biotech-nology Co., Ltd. (Shandong, China). Anti-IMI monoclonal antibody and anti-CLO monoclonal antibody were purchased from Shandong Landu Biotechnology Co., Ltd. (Shandong, China). BCA (bicinchoninic acid) protein quantification kit was purchased from Solarbio (Beijing, China). Monodispersed Magnetite Microspheres (MNPs) were purchased from Beijing Baseline Co., Ltd. (Tianjin, China). The nanoparticle morphology and size of UCNPs were determined by FEI TECNAI G20 transmission electron microscopy (TEM, FEI Company, Hillsboro, OR, USA). Fourier transform infrared spectroscopy (FTIR) of UCNPs was determined by an FTIR spectrophotometer (Perkin Elmer, Waltham, MA, USA). Powder X-ray diffraction (XRD) measurements of UCNPs were performed by an AXIS-ULTRA-DLD instrument (Millipore, New York City, NY, USA). The TEM, FTIR and XRD characterizations of UCNPs were determined by the School of Chemistry, Tianjin Normal University. The fluorescence intensity of the upconverting nanoparticles was detected by an F-2500 spectrofluorometer equipped with a 980 nm laser exciter (Shimadzu Corporation, Kyoto, Japan). The high-performance liquid chromatograph (HPLC) was performed by Thermo Fisher Scientific, Waltham, MA, USA. All chemicals used were analytical grade.

2.2. *Preparation of Upconversion Nanomaterials*

2.2.1. Preparation of $NaYF_4$:Yb@$NaYF_4$:Ho UCNPs

The $NaYF_4$:Yb@$NaYF_4$:Ho core-shell structure upconversion material was prepared based on a high-temperature thermal decomposition method. The total mass of 2 mmol $ReCl_3$ (including 1 mmol $YbCl_3{\cdot}6H_2O$ and 1 mmol $YCl_3{\cdot}6H_2O$), 14 mL of oleic acid and 30 mL of 1-octadecene were used as raw materials. Under the protection of high-purity argon, the temperature was raised to 160 °C and kept at this temperature for 60 min. After the reactant was cooled to room temperature, a methanol solution dissolved with 8 mmol NHF_4 and 5 mmol NaOH was added dropwise. After stirring at room temperature for 30 min, the reactant was heated to 100 °C and incubated for a period of time so that the methanol solution in the reactant could be fully evaporated, then the reactant was heated to 300 °C and kept at this temperature for 60 min, after the reactant cooled to room temperature, it was washed with ethanol and cyclohexane several times and dried for use. The prepared upconversion nanoparticles were used as the core. During the encapsulation process, 0.5 mmol of $ReCl_3$ (including $YCl_3{\cdot}6H_2O$ and $HoCl_3$) was added to the prepared core dissolved in 3 mL of cyclohexane after the reaction was completed at 160 °C for 60 min. The mixture was stirred at room temperature for 20 min, then added to a methanol solution dissolved in 2 mmol NHF_4 and 1.25 mmol NaOH. After stirring at room temperature for 30 min, the reactant was heated to 100 °C and kept for a period of time, so that the methanol solution in the reactant could be fully evaporated, and then the reactant was heated to 300 °C and kept at this temperature for 60 min. After the reactant was cooled to room temperature, it was washed several times with ethanol and cyclohexane, and dried for later use.

2.2.2. Preparation of $NaYF_4$:Yb,Tm UCNPs

The preparation of $NaYF_4$:Yb,Tm was similar to the nuclear layer structure of $NaYF_4$:Yb@$NaYF_4$:Ho. In a total mass of 2 mmol $ReCl_3$ (including 1.6 mmol $YCl_3{\cdot}6H_2O$, 0.36 mmol $YbCl_3{\cdot}6H_2O$ and 0.04 mmol $TmCl_3{\cdot}6H_2O$), 14 mL of oleic acid and 30 mL of 1-octadecenc were added. The 1-octadecene was used as the raw material, and the temperature was raised to 160 °C under the protection of high-purity argon and kept for 60 min. After the reactant was cooled to room temperature, a methanol solution dissolved with 8 mmol NHF_4 and 5 mmol NaOH was added dropwise, and stirred at room temperature for 30 min, the reactant was heated to 100 °C and kept at this temperature for a period of time, so that the methanol solution in the reactant could be fully evaporated, then the reactant was heated to 300 °C and kept at this temperature for 60 min. After this, the reactant was cooled to room temperature, and washed several times with ethanol and cyclohexane before a secondary drying process was used to preserve it for later use.

2.2.3. Preparation of $NaYF_4$:Yb,Ho UCNPs

The synthesis of $NaYF_4$:Yb,Ho UCNPs is similar to the preparation process of $NaYF_4$:Yb,Tm UCNPs, changing the molar ratio of Y:Yb:Ho = 80:18:2.

2.3. Surface Modification of Upconversion Nanomaterials

Surface modification of oil-soluble upconversion materials was accomplished using the classical Stober method. Using sonication, we dispersed 20 mg of oil-soluble UCNPs in 60 mL of isopropanol, add 2.5 mL of ammonia water and 20 mL of water to the flask, and kept the mixture at 35 °C under vigorous stirring. Then, 20 mL of isopropanol solution containing 50 μL of TEOS was added dropwise to the mixture and the reaction was maintained for another 3 h. Then, a solution containing 30 mL of isopropanol and 200 μL of APTES was added dropwise to the suspension for 1 h. After the reaction, the product was aged at room temperature for 2 h, the precipitate was separated by centrifugation, washed three times with deionized water and ethanol, and dried in an oven at 60 °C for 12 h to obtain amino-modified $NaYF_4$:Yb@$NaYF_4$:Ho and $NaYF_4$:Yb,Tm UCNP.

2.4. Preparation of Capture Probes and Signal Probes

The signal probe was prepared using the classic glutaraldehyde cross-linking method. 10 mg of the upconversion material was dissolved in 5 mL of 10 mmol/L PBS (phosphate buffer saline) solution for ultrasonic dispersion for 30 min, and 1.25 mL of 25% glutaraldehyde solution and 100 mg of sodium borohydride were added at room temperature. After the reaction was completed, the precipitate was collected by centrifugation. The solid powder was washed three times with PBS, resuspended in 5 mL of 10 mmol/L PBS for ultrasonic dispersion, and a certain amount of antibody was added to be slowly shaken at room temperature for 6 h. The supernatant was collected by centrifugation at the same time, and the material was washed several times, mixed with 5 mL of 10 mmol/L PBS containing 3% BSA, and slowly shaken for 6 h at room temperature. After separation and washing, it was dispersed in 5 mL of PBS and stored at 4 °C.

Similarly, the capture probe was prepared by dissolving 10 mg Fe_3O_4 magnetic microspheres in 5 mL 10 mmol/L PBS solution. The operation method is similar to the preparation of the above signal probe. A certain amount of antigen was added and finally dispersed in 5 mL PBS buffer to obtain a capture probe with a concentration of 2 mg/mL.

2.5. Establishment of Fluorescence Immunoassay

In a typical experiment, 100 μL of IMI and CLO standards and 100 μL of capture probes were mixed in a 1.5 mL centrifuge tube, then a 200 μL of (each) signal probe was added to the mixture, and left to compete for 60 min. After the reaction, we collected the signal probe-capture probe complex through the external magnetic field, washed it with PBS three times, dispersed the precipitate in 500 μL PBS, and used the fluorescence spectrophotometer with an external 980 nm exciter to measure the fluorescence intensity values at 656 nm and 696 nm, respectively. After optimizing the pH value (5.0, 6.0, 7.4, 8.0, 9.0), Na^+ concentration (10, 20, 30, 40, 50 mmol/L) and methanol concentration (0%, 2.5%, 5%, 10%) of PBS buffer solution in the competitive reaction system, the standard curve was established by changing the competitive reaction time. Finally, the optimal pH value, ion concentration, methanol content and competitive reaction time of the buffer solution were determined by F_{max}/IC_{50} value. (F_{max} is the maximum value of fluorescence intensity, and IC_{50} is the corresponding standard concentration when the inhibition rate is 50%, i.e., sensitivity.)

2.6. Specificity

In order to investigate the specificity of the method, five other structural analogs to IMI and CLO were selected, i.e., acetamiprid, thiacloprid, nitenpyram, dinotefuran, and thiamethoxam. Acetamiprid, thiacloprid, nitenpyram, dinotefuran, thiamethoxam and IMI, CLO all belong to neonicotinoid pesticides and are structural analogs, and their application scope is also similar. Therefore, these pesticides were selected as specific research objects. Their standards were formulated at a concentration of 10 μg/L to cross-react with the capture probe signal probes. The fluorescence intensity values at emission wavelengths of 656 nm and 696 nm were recorded as I values. Only the fluorescence intensity values of the capture probe and the signal probe were recorded as I_0, and $\Delta I = I_0 - I$ was calculated to evaluate the specificity of the method.

2.7. Sample Handling

In order to further evaluate the detection accuracy and practical applicability of IMI and CLO by fluorescence immunoassay, an additive recovery experiment was carried out. The coefficient of variation (CV) was used as the basis for judging the precision of this method and each concentration was repeated 5 times. In this study, six fruit and vegetable samples were purchased from the market, including apples, oranges, peaches, cucumbers, tomatoes, and peppers. The extracted fruit and vegetable samples were pre-treated, 2 g of edible parts (accurate to 0.01 g) were weighed into a 50 mL centrifuge tube, 4 mL acetonitrile was added, and a high-speed tissue masher was used. Then, we homogenized an extract at 15,000 rpm for 1 min, added 1 g of sodium chloride, then homogenized an extract for 1 min, centrifuged at 3000 rpm for 5 min, took 2 mL of the supernatant into a 50 mL round-bottom flask, and rotated at 38 °C to dryness added 1 mL of 25% acetonitrile, sonicate until fully dissolved, and passed through a 0.22 μm filter.

2.8. Recovery Experiment

To perform the recovery experiment, we added 2, 20 and 200 ng/mL IMI and CLO standard solutions to the blank samples of pepper and cucumber, respectively, and added 5, 50 and 500 ng/mL IMI and CLO standard solutions to the blank samples of tomato, apple, orange and peach, respectively. After mixing, we treated them with the above sample treatment methods. Finally, we diluted them with PBS appropriately, and used this experimental method for detection.

2.9. HPLC Comparison Test

The chromatographic conditions of HPLC are as follows: chromatographic column: Bridge C18 (5 μm, 4.6 × 150 mm); mobile phase: methanol-water (32:68); flow rate 1.0 mL/min; column temperature: 30 °C; detection wavelength: 270 nm; injection volume: 20 μL.

The edible part of the fruit and vegetable samples was divided into two parts, and equal amounts of standard substances were added, respectively. One part was subjected to sample pretreatment as described in Section 2.6 and tested by this experimental method; the other part refers to GB/T 20769-2008 [27]. After the sample pretreatment, HPLC was used for detection, that is, after the sample was extracted with acetonitrile, salted out, centrifuged, and rotary evaporated, it was purified by a solid-phase extraction column and eluted with acetonitrile and toluene (3+1). After rotary evaporation, it was dried with nitrogen, and then acetonitrile decahydrate (3+2) solution was quickly added to mix, well-filtered through a 0.22 μm filter membrane, and then detected by HPLC.

2.10. Actual Sample Detection

The samples were selected from the local market (Tianjin, China), three each of peppers, cucumbers, tomatoes, oranges, peaches and apples, a total of 18 blank samples, numbered 1–18#, randomly added IMI and CLO standards to make blind samples, for the detection of immunoassay and HPLC in this experiment.

3. Results and Discussion

3.1. Characterization of the Prepared UCNPs

The morphologies of upconversion nanomaterials with oil-soluble and water-soluble $NaYF_4$:Yb@$NaYF_4$:Ho core-shell structure and $NaYF_4$:Yb, Tm monolayer structure were characterized by transmission electron microscopy. The results are shown in the figure: TEM images show that the two synthesized oil-soluble upconversion materials are hexagonal phases with smooth surfaces, uniform size and good dispersion, among which the NaYF4:Yb@NaYF4:Ho (Figure 1B) core-shell structure particles. $NaYF_4$:Yb UCNPs (Figure 1A) are the core of oil-soluble core-shell $NaYF_4$:Yb@$NaYF_4$:Ho UCNPs. The particle size of $NaYF_4$:Yb UCNPs is about 75 nm and the crystal form is in a hexagonal phase, but partially spherical. A shell containing the rare earth element Ho was coated on the surface of $NaYF_4$: Yb UCNPs to obtain $NaYF_4$:Yb@$NaYF_4$:Ho UCNPs. The particle size of $NaYF_4$:Yb@$NaYF_4$:Ho UCNPs is about 80nm, and the crystal form is in a standard β-hexagonal phase. It can be seen that this paper successfully coated a shell on the surface of $NaYF_4$:Yb UCNPs by thermal decomposition method to form $NaYF_4$:Yb@$NaYF_4$:Ho UCNPs. The particle size of $NaYF_4$:Yb,Tm (Figure 1C) monolayer structure is about 60–70 nm. The surface-modified upconversion materials (Figure 1D,E) are spherical particles with uniform particle size, and the surface is coated with a layer of silicon shell with a thickness of about 10 nm, which has good dispersibility in the water system. Oil-soluble nanoparticles were transformed into water-soluble nanoparticles by coating silicon shells on the surface of UCNPs, and their biocompatibility was improved by modifying amino groups on their surface.

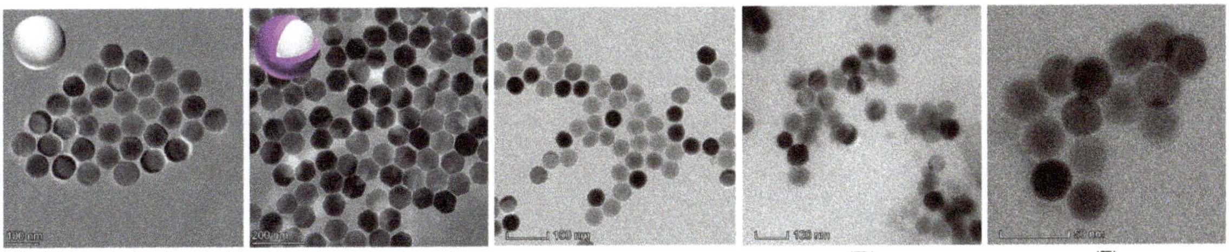

(A) (B) (C) (D) (E)

Figure 1. Transmission electron microscopy imaging of (**A**) oil-soluble core $NaYF_4$:Yb UCNPs; (**B**) oil-soluble core-shell $NaYF_4$:Yb@$NaYF_4$:Ho UCNPs; (**C**) oil-soluble bare core $NaYF_4$:Yb,Tm UCNPs; (**D**) water-soluble core-shell $NaYF_4$:Yb@$NaYF_4$:Ho UCNPs; (**E**) water-soluble naked core structure $NaYF_4$:Yb,Tm UCNPs.

Under an 980 nm excitation light, $NaYF_4$:Yb@$NaYF_4$:Ho core-shell UCNPs produced four emission peaks at 408, 476, 542 and 656 nm, respectively, and the largest characteristic emission peak was at 656 nm (Figure 2A). $NaYF_4$:Yb,Tm upconversion nanomaterials with bare core structure produced three emission peaks at 476, 648 and 696 nm, respectively, and 696 nm was the largest characteristic emission peak (Figure 2B). After mixing the above two materials, the emission peaks at 656 nm and 696 nm are independent and do not overlap, which can realize the simultaneous detection of two targets (Figure 2C). Through horizontal comparison, it is not difficult to find that the fluorescence intensity of core-shell structure is an order of magnitude higher than that of single-layer structure upconversion materials. This shows that core-shell upconversion materials can effectively reduce the quenching effect of surface defects such as hanging bonds and polymer vibrational groups on the luminescence of nanomaterials. It can effectively improve the luminous efficiency of the material, to further enhance the fluorescence. In this paper, Yb/Ho system was studied, and the proportion of doped rare earth elements was optimized in the synthesis of core-shell UCNPs. We adjusted the emission wavelength of $NaYF_4$:Yb@$NaYF_4$:Ho and enhanced its emission light after 600 nm, and widenened the application of this material in the fields of biological imaging, detection and multi-color upconversion.

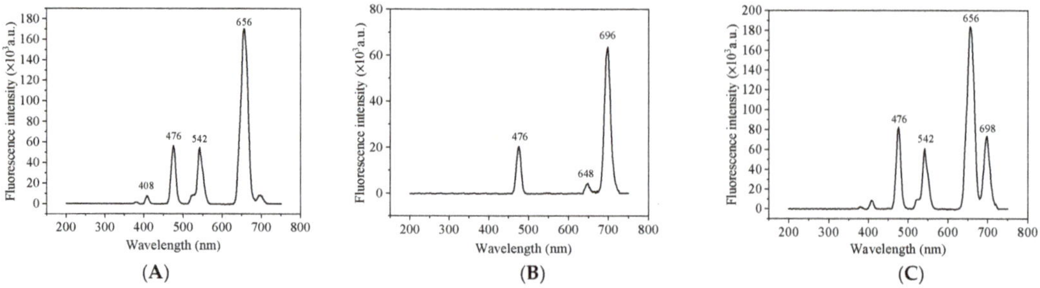

Figure 2. The fluorescence spectrum of (**A**) $NaYF_4$:Yb@$NaYF_4$:Ho; (**B**) $NaYF_4$:Yb,Tm and (**C**) the mixture of $NaYF_4$:Yb@$NaYF_4$:Ho and $NaYF_4$:Yb,Tm UNCPs under 980 nm excitation.

Under 980 nm excitation, the core-shell structure UCNPs $NaYF_4$:Yb@$NaYF_4$:Ho and monolayer UCNPs $NaYF_4$:Yb, Ho have two peaks at 540 nm (green light) and 656 nm (red light). The fluorescence intensity ratios of core-shell structure and monolayer structure UCNPs at 540 nm:656 nm are about 1:3 and 13:1. Therefore, $NaYF_4$:Yb@$NaYF_4$:Ho core-shell UCNPs mainly emit red light, while $NaYF_4$:Yb,Ho single-layer structure UCNPs mainly emit green light (Figure 3). Therefore, it can be shown that accurate upconversion luminescence color regulation can be achieved by rationally distributing the spatial position and relative concentration of activator ions. Separating Ho and Yb in $NaYF_4$:Yb, Ho nanoparticles and Yb:Ho = 50:1 can suppress the luminescence at 540 nm and enhance the luminescence at 656 nm, making the Yb, Ho co-doped $NaYF_4$ change from green to red.

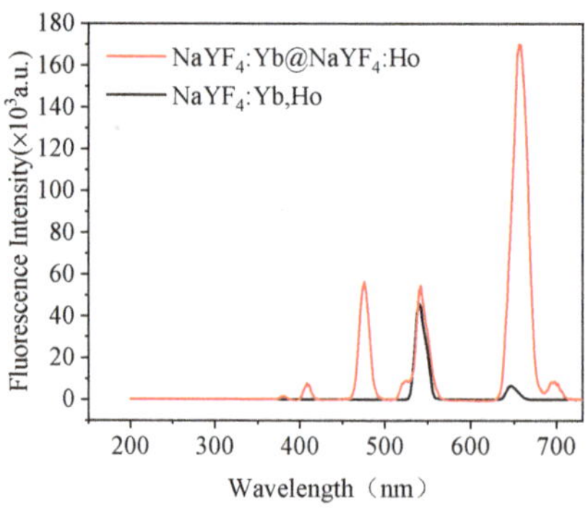

Figure 3. The fluorescence spectrum of $NaYF_4$:Yb@$NaYF_4$:Ho UCNPs and $NaYF_4$:Yb,Ho UCNPs.

The crystal structure and phase purity were determined by powder X-ray diffraction. The $NaYF_4$:Yb@$NaYF_4$:Ho and $NaYF_4$:Yb,Tm upconversion nanomaterials were identified by X-ray diffraction as shown in Figure 4. The positions of all diffraction peaks and the relative intensities match well with the diffraction patterns of standard hexagonal β-phase UCNPs (JCPDS No. 16-0334). There are no diffraction peaks of other impurities indicating that the as-prepared $NaYF_4$:Yb@$NaYF_4$:Ho and $NaYF_4$:Yb,Tm upconversion nanoparticles are well crystallized and exhibit good photoluminescence effectiveness.

For the $NaYF_4$:Yb@$NaYF_4$:Ho UCNPs (Figure 5A), the symmetric stretching vibration of Si-O appeared in the region of about 1100 cm^{-1}, which indicated that the surface of UCNP is coated with a silicon shell. The stretching and bending vibrations of amino groups appear at 3183 cm^{-1}. The peak at 2933 cm^{-1} belongs to the asymmetric and symmetric stretching vibrations of the methylene group. Similar results were also observed in the infrared spectra of $NaYF_4$:Yb,Tm (Figure 5B). For the $NaYF_4$:Yb,Tm UCNPs, the symmetric stretching vibration of Si-O appeared in the region of 1097 cm^{-1}, which shows that the material has been coated with a silicon shell. The stretching and bending vibration of the amino group appears at 3162 cm^{-1}, indicating that the surface of the material has been modified with the amino group. Fourier transform infrared spectroscopy results show that both materials have been coated with silicon shells and modified with amino groups. The modified UCNPs have good water solubility and are conducive to binding with antibodies in immunoassay. The modified UCNPs have good dispersibility in the water system, no significant change in morphology, uniform particle size, and a layer of silicon shell with a thickness of about 5 nm is wrapped on the surface.

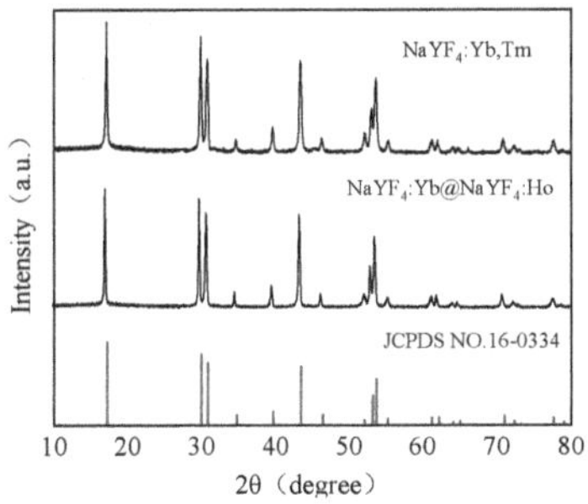

Figure 4. X-ray diffraction patterns of $NaYF_4$:Yb@$NaYF_4$:Ho UCNPs and $NaYF_4$:Yb,Ho UCNPs.

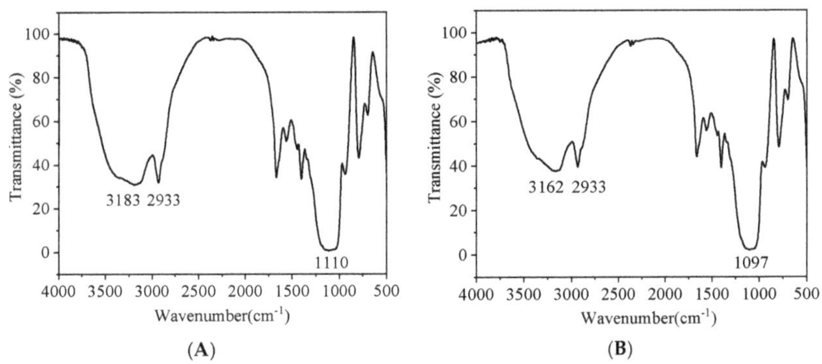

Figure 5. (**A**) Fourier transform infrared spectra of water-soluble $NaYF_4$:Yb@$NaYF_4$:Ho UCNPs; (**B**) and $NaYF_4$:Yb,Tm UCNPs.

3.2. *Optimize Probe Preparation Conditions*

The preparation conditions of signal probes and capture probes directly affect the sensitivity and accuracy of fluorescent immunoassays. The BCA protein concentration assay kit was used to detect the coupling amount of antibody and coated antigen on the surface of UCNPs and MNPs. As shown in Figure 6A, when 5 mL of 2 mg/mL modified water-soluble UCNPs were immobilized, when the amount of antibody added was less than 80 μg, the coupling amount increased with the increase of the amount of antibody added, and the coupling rate decreased with the increase of the added amount. When the addition amount is greater than 80 μg, the coupling amount remains unchanged, and the coupling rate shows a significant downward trend, indicating that the amount of antibody coupled on the surface of the material has reached the saturation state, and excessive addition will cause waste. Therefore, 80 μg of CLO antibody was added to prepare a signal probe, and the coupling rate was 86.4%. At the same time, 5 mL of 2 mg/mL activated MNPs were fixed, and different amounts of the coating were added. According to Figure 6B, 80 μg of CLO antigen was finally added to prepare a capture probe, and the coupling rate was 77.7%. Similarly, the optimal dosage of IMI antibody (Figure 6C) was determined to be

100 μg, and the coupling rate was 72.5%; the optimum dosage of IMI antigen (Figure 6D) was 80 μg, and the coupling rate was 78.2%.

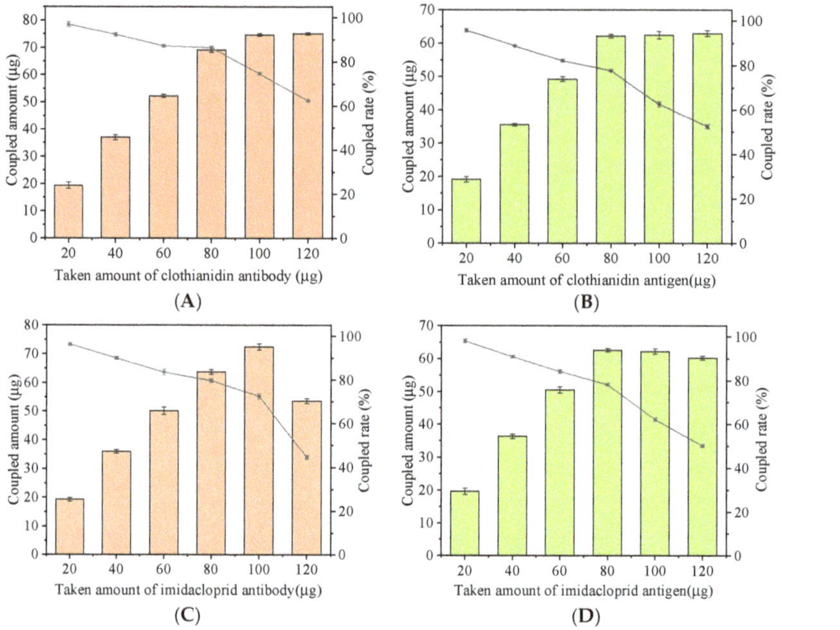

Figure 6. Optimizing the loading of Ab and Ag on signal (pink) and capture (green) probes. Optimizing the loading of (**A**) CLO antibody; (**B**) CLO antigen; (**C**) IMI antibody; (**D**) IMI antigen.

By changing the ratio of the signal probe and capture probe, the addition amount of the capture probe is optimized to obtain the best binding addition amount of antigen and antibody. In this experiment, the addition amount of both signal probes is 200 μL. The volumes of the two capture probes range from 40 to 140 μL. After the reaction, the fluorescence value is measured by a fluorescence spectrophotometer. According to the changing trend of fluorescence value (Figure 7), the fluorescence intensity of the reaction system increases with the increase of the addition of IMI and CLO capture probes. When the addition of both capture probes is 100 μL, the fluorescence intensity reaches the maximum. Increasing the amount of capture probe will lead to the darker color of the reaction system, cause background interference and reduce the fluorescence intensity. Therefore, the best addition amount of IMI and CLO capture probe is 100 μL.

3.3. Optimization of Method Conditions

The IC_{50} of this experimental method is the sensitivity. In order to achieve the best sensitivity, the parameters of the competition system were optimized: (1) pH value of buffer solution (2) Na^+ concentration of buffer solution (3) methanol content. Taking IC_{50} and F_{max}/IC_{50} as the evaluation criteria, the minimum value of IC_{50} and the maximum value of F_{max}/IC_{50} were taken as the optimal conditions.

Under the condition of immune reaction, the sensitivity of immune reaction will be affected by over acid or over alkali environment. Adjust the pH value in the system to 5.0, 6.0, 7.4, 8.0 and 9.0. As shown in Figure 8A,D, IC_{50} first decreased and then increased with the increase of pH value. When the pH value was 7.4, the IC_{50} value was the lowest, while F_{max}/IC_{50} reached the highest value. Therefore, pH 7.4 was selected as the optimal reaction system. The ion concentration in the buffer, i.e., Na^+ concentration, will also interfere with the results. Set the Na^+ concentration to five concentrations: 0.01, 0.02, 0.03, 0.04 and

0.05 mol/L. With the increase of Na^+ concentration in the buffer (Figure 8B,E), the IC_{50} of IMI and CLO showed a trend from decreasing to increasing and then decreasing. When the Na^+ concentration was 0.02 mol/L, IC_{50} reached the lowest, while F_{max}/IC_{50} was the highest, and the Na^+ concentration of the final selection buffer was 0.02 mol/L. In order to improve the solubility of the target in the aqueous phase, a certain amount of methanol was added to the system. Set the methanol content to 0%, 2.5%, 5%, 10%. The results are shown in Figure 8C, F, where IC_{50} increased significantly with the increase of methanol content, and F_{max}/IC_{50} showed a downward trend. When the methanol content was 2.5%, the IC_{50} value reached the lowest value, while the F_{max}/IC_{50} value reached the highest value. The methanol content of the final selection buffer was 2.5%.

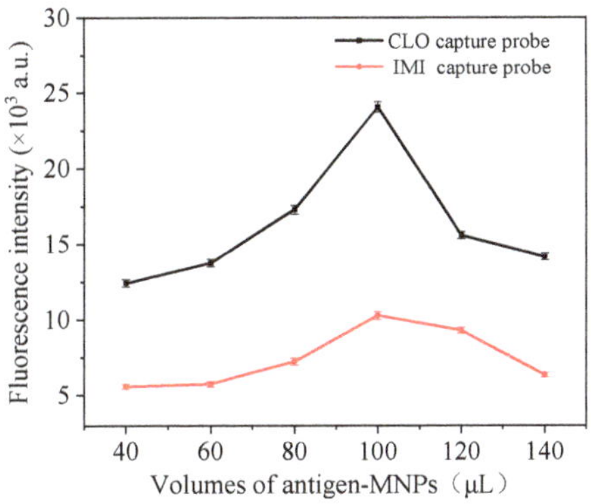

Figure 7. Optimization of the amount of capture probe added.

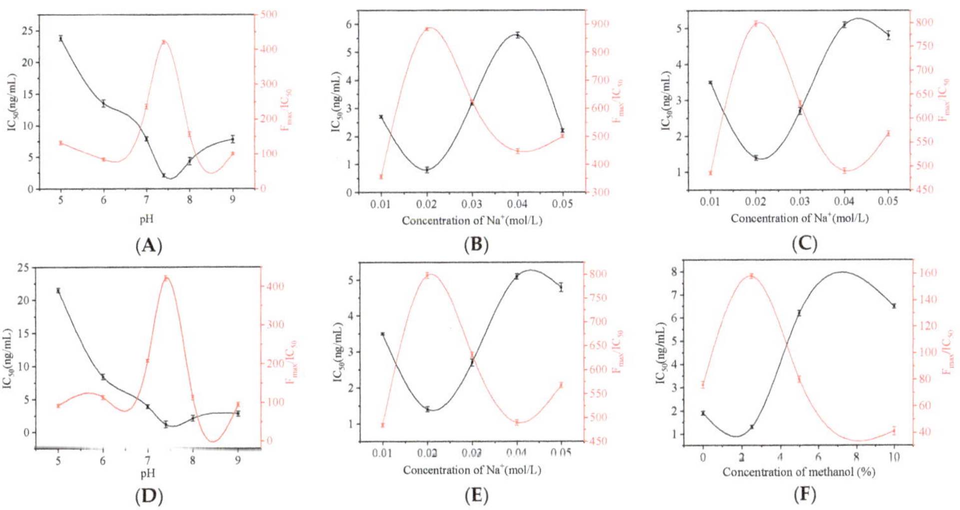

Figure 8. Optimization of pH, Na+ concentration, methanol content of (**A–C**) IMI and (**D–F**) CLO.

3.4. Determination of the Detection Limit of the Method

In this study, the characteristic absorption peak at 656 nm of $NaYF_4$:Yb@$NaYF_4$:Ho UCNPs was used as the detection wavelength of CLO, and the characteristic absorption peak at 696 nm of $NaYF_4$:Yb,Tm UCNPs was used as the detection wavelength of IMI. An upconversion magnetic separation fluorescence immunoassay method for the simultaneous detection of IMI and CLO was established. Under the best competitive reaction conditions, the standard solutions of different concentrations of IMI and CLO were mixed with a 100 μL capture probe in a 1.5 mL centrifuge tube. Then, 200 μL signal probes were added and reacted at room temperature for 1 h. After the reaction, it was separated by the external magnetic field and washed with PBS 3 times. Under the optimal competitive reaction conditions, the change in fluorescence intensity ΔI ($\Delta I = I_0 - I$; I_0 and I represent the fluorescence intensity in the absence and presence of analytes, respectively) and the logarithm of the concentrations of IMI and CLO (Figure 9A). The limit of detection (LOD) and sensitivity (IC_{50}) of IMI was 0.032 ng/mL and 4.7 ng/mL, respectively, and the linear range (IC_{10}–IC_{90}) was 0.032–285.75 ng/mL (Figure 9B). The LOD and the IC_{50} of CLO were 0.028 ng/mL and 2.1 ng/mL, and the linear range was 0.028–200 ng/mL (Figure 9C). Tao Zhexuan [28] established a multicolor upconversion immunoassay method for the detection of IMI and thiacloprid by synthesizing $NaYF_4$:Yb,Er UCNPs and $NaYF_4$:Yb,Tm UCNPs. The LOD of IMI was 0.32 ng/mL and IC_{50} was 5.8 ng/mL; LOD of thiacloprid was 0.61 ng/mL and IC_{50} was 6.45 ng/mL. Compared with Tao's [28] experimental method, he synthesized $NaYF_4$:Yb,Er UCNPs and $NaYF_4$:Yb,Tm UCNPs by hydrothermal method, this experiment synthesized $NaYF_4$:Yb,Tm UCNPs and $NaYF_4$:Yb@$NaYF_4$:Ho UCNPs have higher fluorescence intensity by thermal decomposition method. Applying these two UCNPs to the upconversion immunoassay method for the simultaneous detection of IMI and CLO has the advantages of lower detection limit and higher sensitivity, which can provide technical support for the detection of neonicotinoids pesticides.

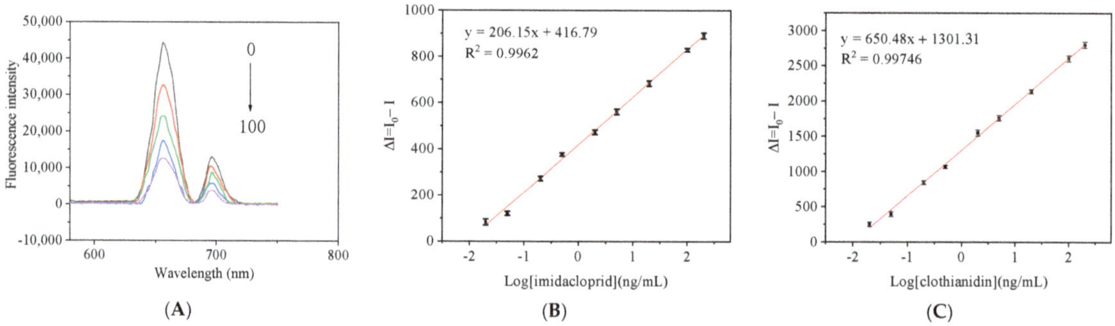

Figure 9. (**A**) Upconversion fluorescence spectra of UCNPs as a function of IMI or CLO concentrations; linear relationship between upconverted fluorescence intensity with (**B**) IMI and (**C**) CLO concentration.

3.5. Specificity

In order to evaluate the specificity of the fluorescence immunoassay method, 10 μg/L IMI, CLO and their structural analogs were added to the reaction system for detection, and the corresponding fluorescence intensity difference was calculated according to the blank fluorescence intensity (Figure 10). When IMI and CLO were added, the fluorescence changed significantly, indicating that the UCNPs were bound with the antibody specifically binding to IMI and CLO. When the ACE, THI, NIT, DIN, THX and other neonicotinoids pesticide structural analogs were added, it only resulted in some nonspecific adsorption. There was almost no specific reaction with the antibody on the UCNPs. Therefore, the results show that this method has high specificity.

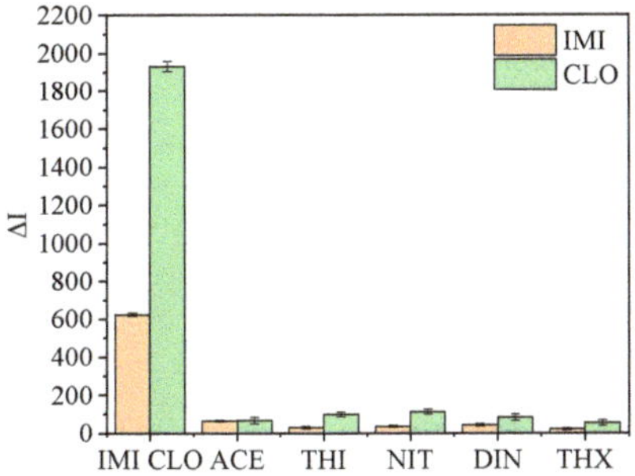

Figure 10. Method specificity evaluation (The concentrations of all standards are 10 μg/L).

3.6. Elimination of Matrix Effects

The vegetable and fruit samples used in this experiment will cause certain interference to the detection results due to the influence of sugars, pigments and other matrices. After extracting the six vegetable and fruit samples of apple, orange, peach, cucumber, tomato and pepper with acetonitrile, the method of buffer dilution was used to eliminate the influence of the matrix. It can be seen from Figure 11 that the curves for cucumber and pepper diluted with PBS 10 times and apple, orange, peach and tomato diluted with PBS 20 times basically coincide with the standard curve, indicating that this method can effectively eliminate matrix interference.

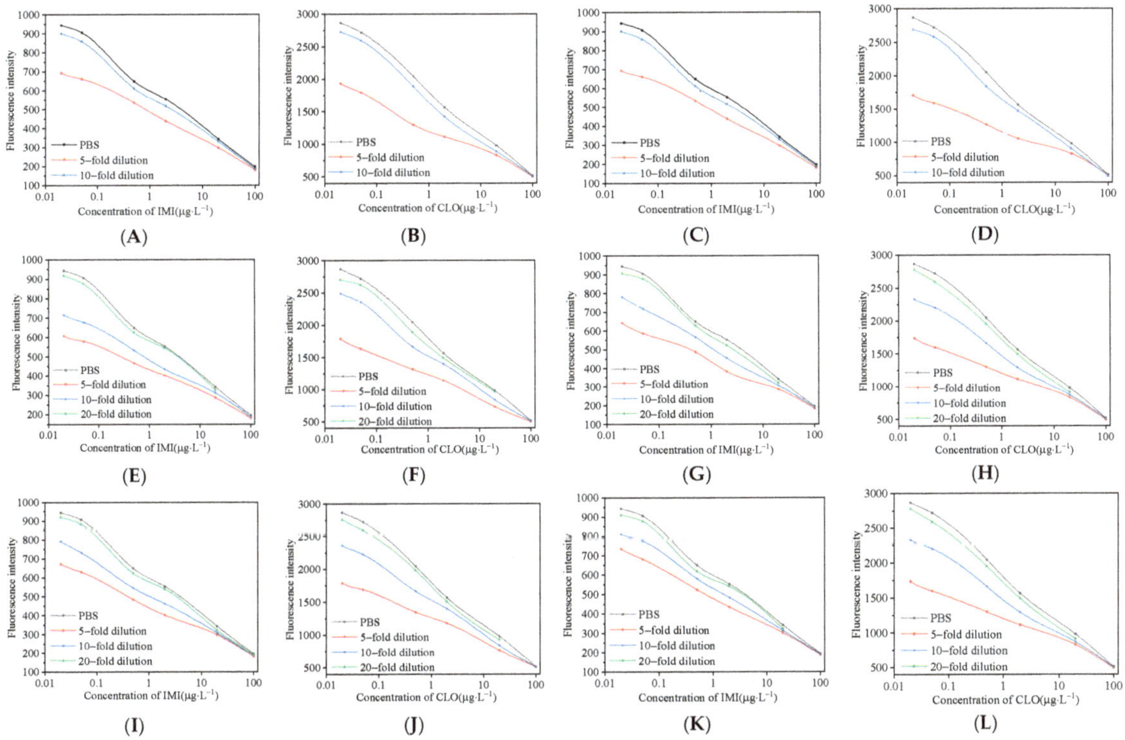

Figure 11. Evaluation of matrix effect in fruit and vegetable acetonitrile extracts; (**A**,**B**) cucumber; (**C**,**D**) pepper; (**E**,**F**) apple; (**G**,**H**) orange; (**I**,**J**) peach; (**K**,**L**) tomato.

3.7. Spiked Recovery Experimental Results

IMI and CLO standards with low, medium and high concentrations were added to apple, orange, peach, cucumber, tomato, and pepper, respectively, and five parallel groups were set for the determination of each concentration of each sample. As shown in Table 1, the recoveries of IMI and CLO were 83.33–115.02% and 84.49–106.64%, respectively, and the coefficient of variation was less than 10% for both. In apple, orange, peach, cucumber, tomato and pepper samples, the detection limit of this method was 5 ng/mL, 2 ng/mL, 5 ng/mL, 5 ng/mL, 5 ng/mL, and 5 ng/mL, respectively, which were much lower than those in China Maximum Residue Limit Standards.

Table 1. Average recoveries of IMI and CLO spiked standards.

Sample	Scalar (μg/kg)	IMI			CLO		
		Mean ± SD (μg/kg)	Recovery Rate (%)	Coefficient of Variation (%)	Mean ± SD (μg/kg)	Recovery Rate (%)	Coefficient of Variation (%)
pepper	2	1.71 ± 0.15	85.34	8.6	1.99 ± 0.18	99.28	9.0
	20	20.72 ± 1.43	103.58	6.9	21.33 ± 1.30	106.64	6.1
	200	182.50 ± 7.85	91.25	4.3	168.98 ± 6.08	84.49	3.6
cucumber	2	1.90 ± 0.17	95.10	9.2	1.93 ± 0.17	96.27	8.9
	20	17.98 ± 0.91	89.89	5.1	20.55 ± 0.97	102.75	4.7
	200	180.47 ± 3.43	90.24	1.9	175.94 ± 4.57	87.97	2.6
tomato	5	4.29 ± 0.35	85.74	8.1	5.05 ± 0.35	100.93	6.9
	50	47.11 ± 2.12	94.23	4.5	43.62 ± 1.57	87.24	3.6
	500	480.43 ± 11.53	96.09	2.4	444.02 ± 7.99	88.80	1.8
orange	5	4.17 ± 0.30	83.33	7.2	4.65 ± 0.35	92.94	7.5
	50	50.15 ± 2.81	100.31	5.6	50.38 ± 1.66	100.76	3.3
	500	425.36 ± 13.19	85.07	3.1	476.60 ± 7.63	95.32	1.6
peach	5	4.31 ± 0.29	86.12	6.8	4.37 ± 0.27	87.48	6.2
	50	47.96 ± 1.97	95.92	4.1	49.18 ± 1.67	98.36	3.4
	500	570.60 ± 15.41	114.12	2.7	489.07 ± 9.78	97.81	2.0
apple	5	4.40 ± 0.33	87.97	7.5	4.37 ± 0.34	87.36	7.7
	50	55.96 ± 2.80	111.91	5.0	49.69 ± 1.89	99.38	3.8
	500	575.08 ± 19.55	115.02	3.4	522.91 ± 6.27	104.58	1.2

3.8. Correlation of Immunoassays with HPLC

The concentrations of IMI and CLO in cucumber were determined using this experimental immunoassay and HPLC. As shown in Figure 12, the correlation equations for IMI and CLO are $y = 0.97988x + 1.03381$ ($R^2 = 0.9962$) and $y = 1.00135x - 0.06768$ ($R^2 = 0.9986$), respectively. The results show a good correlation between the two methods, and the immunoassay in this experiment is accurate and reliable and can be used to detect IMI and CLO in real samples. The retention times of IMI and CLO were 11.5 and 18 min, respectively.

3.9. Test Results of Actual Samples

A total of 18 blank samples of 3 peppers, cucumbers, tomatoes, oranges, peaches and apples, were numbered 1–18#. They were randomly added with IMI and CLO standards to make blind samples for immunoassay and HPLC methods. The results are shown in Table 2: Six IMI positive samples were finally detected, including one pepper, one cucumber, one tomato, one orange, one peach and one apple. There were 6 CLO positive samples, including 2 cucumbers, 1 orange, 1 peach and 2 apples. In order to further evaluate the detection accuracy and practical applicability of the fluorescent immunoassay method in this experiment, the detection of IMI and CLO in actual samples was carried out. The results show that in Table 2, the detection results of the established fluorescence immunoassay method are in good agreement with the HPLC detection results. This result can be used for the determination of actual samples.

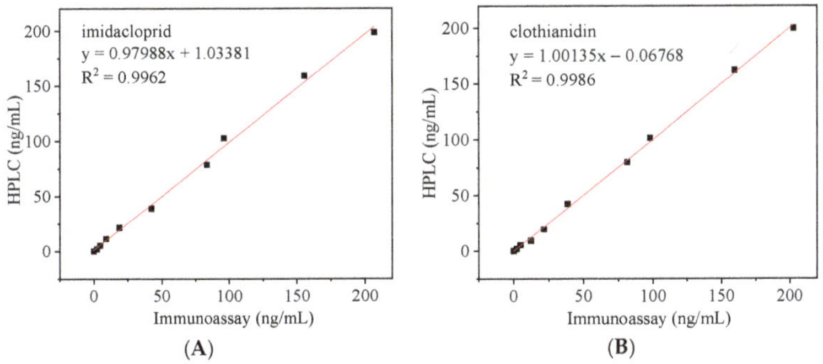

Figure 12. Correlation between HPLC and the method of this study for the detection of (**A**) IMI and (**B**) CLO.

Table 2. Detection of IMI and CLO in actual samples.

Sample Serial Number	This Experimental Method		HPLC	
	IMI (ng/mL)	CLO (ng/mL)	IMI (ng/mL)	CLO (ng/mL)
1#pepper	-	-	-	-
2#pepper	-	-	-	-
3#pepper	22.7	-	23.2	-
4#cucumber	-	82.4	-	85.2
5#cucumber	-	-	-	-
6#cucumber	61.8	103.4	63.7	101.1
7#tomato	92.1	-	94.8	-
8#tomato	-	-	-	-
9#tomato	-	-	-	-
10#orange	-	19.2	-	20.3
11#orange	136.4	-	132.8	-
12#orange	-	-	-	-
13#peach	31.6	54.7	30.2	55.8
14#peach	-	-	-	-
15#peach	-	-	-	-
16#apple	-	36.7	-	38.1
17#apple	48.7	-	47.9	-
18#apple	-	112.3	-	114.1

-: means not detected.

4. Conclusions

In this paper, a novel core-shell structure material is applied to the analysis of immune detection for the first time. Based on the different emission spectra of Yb/Ho and Yb/Tm doped $NaYF_4$ UCNPs excited by a 980 nm laser, a dual-signal probe is prepared. The magnetic nanoparticle trapping probe can realize the principle of rapid separation of immune complexes from the test system. An upconversion-magnetic separation fluorescence immunoassay was successfully developed for simultaneous detection of IMI and CLO. This method has the advantage of causing a rapid, simple, sensitive, innovative transformation of the core-shell nanoparticles coupling antibody, compared to a traditional transformation of materials with higher luminous efficiency and gain better detection results in fluorescence immunoassay detection, proving that the material in the field of food safety testing particularly multicolor code has very strong potential applications. These results provide a new idea for the application of upconversion materials in immunofluorescence in the future.

Author Contributions: Conceptualization, Y.S.; methodology, W.Z. and L.Z.; software, W.Z.; validation, W.Z.; formal analysis, J.J.; investigation, L.Z.; resources, Y.S.; data curation, L.Z.; writing—original draft preparation, W.Z. and L.Z.; writing—review and editing, Y.S.; visualization, J.J.; supervision, Y.S.; project administration, W.Z.; funding acquisition, Y.S. All authors have read and agreed to the published version of the manuscript.

Funding: This research was funded by Open Project Program of State Key Laboratory of Food Nutrition and Safety, Tianjin University of Science and Technology (No. SKLFNS-KF-391 202115) and the Natural Science Foundation of Tianjin (18JCQNJC84400) and National Natural Science Foundation (No. 31301487).

Institutional Review Board Statement: Not applicable.

Informed Consent Statement: Not applicable.

Data Availability Statement: Data is contained within the article.

Acknowledgments: Thanks to Mingfei Pan for his help in this research.

Conflicts of Interest: The authors declare no conflict of interest.

References

1. Jimenez-Lopez, J.; Llorent-Martinez, E.J.; Ortega-Barrales, P.; Ruiz-Medina, A. Analysis of neonicotinoid pesticides in the agri-food sector: A critical assessment of the state of the art. *Appl. Spectrosc. Rev.* **2020**, *55*, 613–646. [CrossRef]
2. Peter, J.; Ralf, N.; Michael, S.; Alfred, E. Overview of the status and global strategy for neonicotinoids. *J. Agric. Food Chem.* **2011**, *59*, 2897–2908.
3. Van der Sluijs, J.P.; Simon-Delso, N.; Goulson, D.; Maxim, L.; Bonmatin, J.M.; Belzunces, L.P. Neonicotinoids, bee disorders and the sustainability of pollinator services. *Curr. Opin. Environ. Sustain.* **2013**, *5*, 293–305. [CrossRef]
4. Sultana, T.; Murray, C.; Kleywegt, S.; Metcalfe, C.D. Neonicotinoid pesticides in drinking water in agricultural regions of southern Ontario, Canada. *Chemosphere* **2018**, *202*, 506–513. [CrossRef]
5. Longing, S.D.; Peterson, E.M.; Jewett, C.T.; Rendon, B.M.; Discua, S.A.; Wooten, K.J.; Subbiah, S.; Smith, P.N.; McIntyre, N.E. Exposure of Foraging Bees (Hymenoptera) to Neonicotinoids in the U.S. Southern High Plains. *Environ. Entomol.* **2020**, *49*, 528–535. [CrossRef]
6. Bishop, C.A.; Woundneh, M.B.; Maisonneuve, F.; Common, J.; Elliott, J.E.; Moran, A.J. Determination of neonicotinoids and butenolide residues in avian and insect pollinators and their ambient environment in Western Canada (2017, 2018). *Sci. Total Environ.* **2020**, *737*, 139386–139429. [CrossRef]
7. Maria, P.; Alzbeta, S.; Valbona, A.; Caterina, F. Impact of Neonicotinoids to Aquatic Invertebrates—In Vitro Studies on *Mytilus galloprovincialis*: A Review. *J. Mar. Sci. Eng.* **2020**, *8*, 801.
8. Chen, Z.L.; Dong, F.S.; Ren, X.; Wu, X.J.; Yuan, L.F.; Li, L.; Li, W.; Zheng, Y.Q. Enantioselective fate of dinotefuran from tomato cultivation to home canning for refining dietary exposure. *J. Hazard. Mater.* **2021**, *405*, 124254. [CrossRef]
9. Mikolic, A.; Karaconji, I.B. Imidacloprid as reproductive toxicant and endocrine disruptor: Investigations in laboratory animals. *Arch. Ind. Hyg. Toxicol.* **2018**, *69*, 103. [CrossRef]
10. Su, Y.Y.; Huang, L.M.; Xiao, Y.Z.; Chao, L.; Jia, W.; Fan, L.Y.; Wang, S.; Zhao, Y.H. Bio-uptake, tissue distribution and metabolism of a neonicotinoid insecticide clothianidin in zebrafish. *Environ. Pollut.* **2021**, *292*, 118317.
11. GB 2763-2021. National Food Safety Standard Maximum Residue Limit of Pesticides in Food. Available online: http://down.foodmate.net/wap/index.php?itemid=97819&moduleid=23 (accessed on 31 March 2022).
12. Himdata, A.; Maria, A.; Alba, B.; Daniela, B.; Luis, C.C.; Chloe, D.L.; Lucien, F.; Luna, G.; Samira, J.; Dimitra, K.; et al. Review of the existing maximum residue levels for imidacloprid according to Article 12 of Regulation (EC) No 396/2005. *Eur. Food Saf. Auth.* **2019**, *17*, 5570.
13. European Food Safety Authority. Reasoned opinion on the review of the existing maximum residue levels (MRLs) for clothianidin and thiamethoxam according to Article 12 of Regulation (EC) No 396/2005. *EFSA J.* **2014**, *12*, 3918.
14. Erickson, B. Neonicotinoids can stay in US market. *Chem. Eng. News* **2020**, *98*, 17. [CrossRef]
15. Science—Analytical Science. Studies from Agriculture and Agri-Food Canada Update Current Data on Analytical Science (High-Throughput Quantitation of Neonicotinoids in Lyophilized Surface Water by LC-APCI-MS/MS). *Sci. Lett.* **2018**. Available online: https://schlr.cnki.net/en/Detail/index/GARJ2018/SPQD29D2BA7ABBA867452A99D741578592D8 (accessed on 31 March 2022).
16. Hem, L.; Abd El-Aty, A.M.; Park, J.H.; Shim, J.H. Determination of dinotefuran in pepper using liquid chromatography: Contribution to safety evaluation. *J. Korean Soc. Appl. Biol. Chem.* **2012**, *55*, 765–768. [CrossRef]
17. Rok, T.; David, H.; Ester, H.; Jernej, M.; Andreja, K.B.; Helena, P. Determination of Neonicotinoid Pesticides in Propolis with Liquid Chromatography Coupled to Tandem Mass Spectrometry. *Molecules* **2020**, *25*, 5870.
18. Watanabe, E.K.; Miyake, S. Direct determination of neonicotinoid insecticides in an analytically challenging crop such as Chinese chives using selective ELISAs. *J. Environ. Sci. Health* **2018**, *53*, 707–712. [CrossRef]
19. Li, D.P.; Gao, Y.J.; Ren, S.M.; Gao, H. Detection of pesticide residues in Honeysuckle by QEChERS extraction and LC-MS/MS technology. *Guangdong Chem. Ind.* **2021**, *48*, 245–248.

20. Hua, X.D.; You, H.J.; Luo, P.W.; Tao, Z.X.; Chen, H.; Liu, F.Q.; Wang, M.H. Upconversion fluorescence immunoassay for imidaclothiz by magnetic nanoparticle separation. *Anal. Bioanal. Chem.* **2017**, *409*, 6885–6892. [CrossRef]
21. Kang, D.K.; Jeon, E.; Kim, S.; Lee, J. Lanthanide-Doped Upconversion Nanomaterials: Recent Advances and Applications. *BioChip J.* **2020**, *14*, 124–135. [CrossRef]
22. Zheng, X.; Kumar, K.R.; Liu, C.G.; Wang, S.B.; Chen, A.Z.; Zhang, Y. Lanthanides-doped near-infrared active upconversion nanocrystals: Upconversion mechanisms and synthesis. *Coord. Chem. Rev.* **2021**, *438*, 17. [CrossRef]
23. Damien, H.; Howard, I.A.; Popescu, R.; Dagmar, G.; Richards, B.S. Structure-Property Relationships in Lanthanide-Doped Upconverting Nanocrystals: Recent Advances in Understanding Core-Shell Structures. *Adv. Mater.* **2019**, *31*, 25.
24. Ye, S.; Chen, G.Y.; Shao, W.; Qu, J.L.; Prasad, P.N. Tuning upconversion through a sensitizer/activator-isolated $NaYF_4$ core/shell structure. *Nanoscale* **2015**, *7*, 3976–3984. [CrossRef]
25. Ye, S.; Song, J.; Wang, D.; Tian, Y.L.; Qu, J.L.; Niu, H.B. Reduced photon quenching in Ce-doped $NaYF_4$:Yb/Ho upconversion nanoparticles with core/shell structure. *Chin. Opt. Lett.* **2016**, *14*, 67–71.
26. Zhang, R. Two Color Upconversion $NaYF_4$:Yb^{3+}/Ln^{3+} (Ln = Er, Tm) Nanorods and Their Anti-Counterfeiting Applications. Master's Thesis, Harbin Institute of Technology, Harbin, China, 2021.
27. GBT 20769-2008. Determination of 450 Pesticides and Related Chemicals Residues in Fruits and Vegetables—LC-MS-MS Method. Available online: https://max.book118.com/html/2019/0107/7052156036002000.shtm (accessed on 31 March 2022).
28. Tao, Z.X. Study on Immunoassay of Imidacloprid and Thiacloprid Based on Nano Materials. Master's Thesis, Nanjing Agricultural University, Nanjing, China, 2019.

foods

MDPI

Article

Construction of an ECL Detection Platform for Sensitive Detection of Carbaryl Based on an Eu^{3+}-Functionalized Metal–Organic Framework Encapsulated with Nanogold

Chang Liu, Haiyang Wang, Xuelian Hu, Yichuan Cao and Guozhen Fang *

State Key Laboratory of Food Nutrition and Safety, College of Food Science and Engineering, Tianjin University of Science and Technology, Tianjin 300457, China; liuchang2022666@163.com (C.L.); ocean5416@163.com (H.W.); hxl523686206@126.com (X.H.); cycydx2019@126.com (Y.C.)

* Correspondence: fangguozhen@tust.edu.cn; Tel.: +86-22-60912493

Abstract: In this work, an Eu^{3+}-MOF-253@Au electrochemiluminescence sensor was successfully constructed for the first time by encapsulating nanogold in the metal–organic frameworks (MOFs) backbone and pore channels, and assembling Eu^{3+} on the MOF backbone. Firstly, the introduction of nanogold overcomes the weakness of MOFs, which was difficult to achieve, and enhances its catalytic performance, followed by the modification of Eu^{3+} to confer the electrochemiluminescence performance and the function of target detection on the sensor. Moreover, carbaryl was placed in an alkaline working solution to enhance the intensity of electrochemiluminescence signal, as well as to promote the hydrolysis of carbaryl into 1-naphthol, which caused the burst of Eu^{3+}-MOF-253@Au electrochemiluminescence sensor, thereby achieving the sensitive detection of carbaryl. On this basis, the electrochemiluminescence detection conditions were optimized, the performance was analyzed, and finally it was successfully used for the detection of carbaryl with good linearity in the range of 0.2–200 μg L^{-1} and a low detection limit (0.14 μg L^{-1}).

Keywords: electrochemiluminescence; MOFs; detection; Eu^{3+}; carbaryl

Citation: Liu, C.; Wang, H.; Hu, X.; Cao, Y.; Fang, G. Construction of an ECL Detection Platform for Sensitive Detection of Carbaryl Based on an Eu^{3+}-Functionalized Metal–Organic Framework Encapsulated with Nanogold. *Foods* **2022**, *11*, 1487. https://doi.org/10.3390/foods11101487

Academic Editor: Olga Gortzi

Received: 5 April 2022
Accepted: 17 May 2022
Published: 19 May 2022

1. Introduction

As an emerging technology in recent decades, electrochemiluminescence (ECL) has been rapidly applied in the analytical fields of environment [1], food [2], clinical [3], and biology studies [4]. Due to the advantages of high sensitivity, high selectivity, fast response, good temporal and spatial control, and simplified setup [5], it has quickly developed into a powerful analysis tool after its discovery around one hundred years ago [6]. Since the first detailed studies in the mid-1960s, the ECL emitter, which is vital in ECL analysis, has been studied and reported on in detail, ranging from the very basic beginning tris(2,2″-bipyridine)ruthenium(II) chloride hexahydrate ($Ru(bpy)_3{}^{2+}$) [7] and classic luminol [8], to the booming quantum dots [9], nanocrystals [10], metal nanoclusters [11], and carbon nitride nanosheets [12]. Admittedly, these electrode materials have their own advantages, such as excellent luminescence signals and high sensitivity under certain conditions, but they also have corresponding weaknesses, such as environmental or biological toxicity, as well as poor stability [13]. Therefore, it is of great significance to explore novel electrode materials with remarkably good performance.

A metal–organic framework (MOF) assembled with metal ions and organic ligands under the action of coordination bonds is a crystalline porous material with an adjustable pore structure, a large specific surface area, and open metal sites [14]. In view of its miraculous performance, it has already been successfully applied in the fields of gas adsorption and storage [15], drug transportation [16], and catalysis [17]. On this basis, MOFs with fluorescence characteristics have gradually emerged in the area of detection [18]. However, the organic branches of MOFs are intricate and complicated, which hinders the transfer of electrons in their structure, and makes them poor in terms of electrical conductivity, thereby limiting its application in the electrochemical sensor field. In order to break this barrier, there are two trains of thought: introducing conductive components into the structure of MOFs by inserting metal nanoparticles or carbon-based nanomaterials into the skeleton, and coordinating the functional molecular composition on the skeleton [19]. Alternatively, according to the synthetic tunability and structural regularity of MOFs, electroactive ligands, such as metalloporphyrin and ruthenium complexes, can be used to construct electroactive MOFs [20]. This certainly broadens the application prospects of MOFs significantly.

Carbaryl is a broad-spectrum insecticide with excellent performance, which was discovered in 1956, and was immediately widely used after it was put on the market. At the end of the 1970s, organo-chlorine pesticides were banned, and the usage of pyrethroids, organophosphorus, and carbamate pesticides was increasing year by year. Among them, carbaryl was also widely used as the first carbamate pesticide to be successfully developed. However, a large amount of carbaryl can cause damage to the human body [21], since excessive carbaryl in the environment will enter poultry, livestock, and even the human body through the respiratory tract, skin mucous membrane, or digestive tract, causing damage to the nervous system, muscles, liver, pancreas, and brain as an acetylcholinesterase inhibitor [22]. This means that it is urgent and significant to establish analytical methods for the detection of carbaryl pesticide residues. Therefore, there were also multiple methods for the detection of carbaryl, such as high performance liquid chromatography(HPLC) [23], mass spectrometry(MS) [24], electrochemical method [25], fluorescence method [26], Raman spectroscopy [27], and colorimetry [28]. These detection methods have achieved satisfactory detection results under certain conditions, but there are also some areas that can be improved, such as improving detection efficiency, improving stability, and lowering detection limits.

In this work, nano-gold with conductive and catalytic properties was introduced onto the surface of the MOFs skeleton, which was loaded with luminescent substances, thus establishing an ECL detection platform. Thereinto, europium, a lanthanide metal, has shown luminescence sensitization and antenna effect in fluorescence detection due to its special chemical properties [18]; it has also been found to produce weak emissions at the cathode in ECL detection [29,30]. In addition, when $K_2S_2O_8$ is used as a co-reactant, ECL can be greatly enhanced [31]. Furthermore, Eu^{3+} was loaded onto the skeleton, and then fixed onto the electrode, which prevented the luminous substance from being dispersed in the working fluid, thereby effectively improving the luminous efficiency and intensity. On the other hand, Eu^{3+} showed a special recognition phenomenon for 1-naphthol, the hydrolysate of carbaryl, in fluorescence detection [32]. Furthermore, the incorporation of nanogold with conductive and catalytic properties into the system can overcome the disadvantage of poor conductivity of MOFs on the one hand, and promote the hydrolysis of carbaryl on the other hand [17]. To go the extra mile, the working solution was adjusted to alkaline to promote the hydrolysis of carbaryl even further, which can also facilitate the occurrence of ECL [33]. These elements of this ECL system complemented each other, and constituted a sensitive and specific electrochemical platform for carbaryl detection.

2. Materials and Methods

2.1. Reagents

Specific reagents and materials are recorded in the Supplementary Materials.

2.2. ECL Measurement

A three-electrode system with a glassy carbon electrode (GCE, 4 mm in diameter) as a working electrode, a platinum disc electrode as a counter electrode, and an Ag/AgCl electrode as a reference electrode was used for the ECL determinations by using an MPI-E II ECL analyzer (Remax Instruments Co., Ltd., Xi'an, China). The electrodes were manufactured by Gaoss Union, Wuhan, China. The ECL measurement was performed in

5 mL of PBS buffer (0.1 M, pH 9) containing 0.05 M $K_2S_2O_8$ with a scan potential range of 0 V to −1.8 V, a scan rate of 0.2 V s^{-1}, and a photomultiplier voltage of 700 V. Furthermore, the cyclic voltammetry (CV) and ECL were recorded simultaneously.

Other instruments and equipment used are recorded in the Supplementary Materials.

2.3. Preparation of Eu^{3+}-Au@MOF-253 Modified Electrodes

The synthesis of MOF-253 was conducted as reported in previous reports, with some modifications [34,35]: $AlCl_3 \cdot 6H_2O$ (151 mg), glacial acetic acid (859 μL), and 2,2′-bipyridine-5,5′-dicarboxylic acid (153 mg) were added to 9 mL of DMF, and reacted in a reactor at 130 °C for 24 h. Then, the powder was washed three times with DMF, and dried under vacuum conditions at 60 °C for 6 h. The resulting product was washed with methanol for 6 h by Soxhlet extraction, and dried at 60 °C for 6 h under vacuum conditions.

MOF-253@Au were synthesized according to the previous method, with some modifications [36]: MOF-253 (100 mg) was dispersed into 125 mL of anhydrous ethanol containing 3 mL of chloroauric acid solution (25 mmol L^{-1}, containing 30 mg $HAuCl_4 \cdot 3H_2O$). Then, 10 mL of aqueous $NaBH_4$ (10 g L^{-1}) was added dropwise, and stirred at room temperature for 1 h. The purple product, MOF-253 encapsulated with nanogold, was thus obtained, and washed three times with methanol, then vacuumed overnight.

The synthesized Eu^{3+}-MOF-253@Au [37]: MOF-253@Au (100 mg) was dispersed into $EuCl_3 \cdot 6H_2O$ (10 mL, 2 mmol) in anhydrous ethanol, and stirred at room temperature for 24 h. The resulting liquid was then separated by centrifugation, washed three times with anhydrous ethanol to remove the excess Eu^{3+}, and finally dried under vacuum conditions at 80 °C for 6 h.

The preparation of Eu^{3+}-MOF-253@Au modified electrodes: Eu^{3+}-MOF-253@Au (8 mg) was added to 250 μL of anhydrous ethanol and 50 μL of Nafion mixed solution, then was sonicated, and dispersed for 30 min. A 5 μL of mixture was then applied dropwise to the surface of the pretreated GCE and dried at room temperature to obtain the modified Eu^{3+}-MOF-253@Au/GCE.

2.4. Sample Preparation

Milk and soybean oil were purchased from the local market. Specific steps for sample processing are described in the Supplementary Materials [38,39].

3. Results and Discussion

3.1. Preparation of Electrodes and Detection of Carbaryl

As exhibited in Scheme 1, after the synthesis of MOF-253, nanogold was encapsulated on its backbone to enhance its electrical conductivity. Then, Eu^{3+} was immobilized on the open chelation site of MOF-253 as the ECL emitter. In order to analyze whether the strong reducing agent $NaBH_4$ had any effect on the structure of MOF-253 when reducing chloroauric acid, scanning electron microscopy (SEM), Fourier transform infrared (FT-IR), and X-ray diffraction (XRD) analyses were performed on MOF-253 and MOF-253@Au, respectively. The patterns are shown in Supplementary Materials.

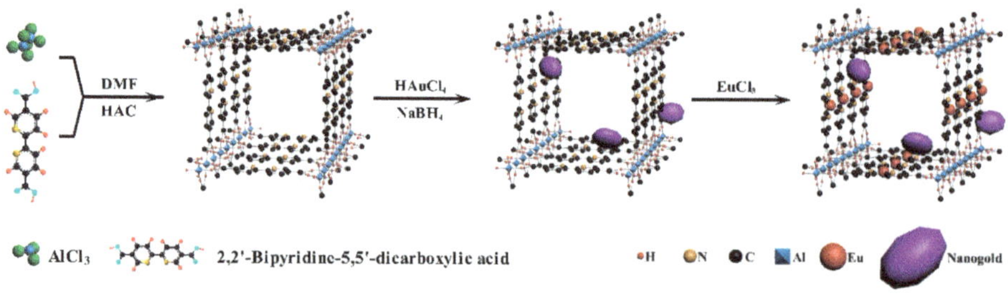

Scheme 1. Schematic diagram of the synthesis of Eu^{3+}-MOF-253@Au.

The material was drop-cast onto the surface of GCE for ECL analysis, as shown in Figure 1a. The bare electrode (black) only showed a weak ECL signal, while the MOF-253 modified electrode (red) showed a much weaker ECL, which was related to the fact that the MOF-253 itself was non-conductive. The blue curve showed an enhanced ECL signal for the MOF-253@Au/GCE, which should be attributed to the fact that the conductivity of nanogold encapsulated in MOF-253 increased the electron transfer capability of the electrode surface, thus enhancing the ECL signal. The Eu^{3+}-MOF-253@Au/GCE (green curve) showed a significant enhancement of the ECL signal, almost four times that of the bare electrode, and twice that of the MOF-253@Au/GCE, which meant that the prepared Eu^{3+}-MOF-253@Au/GCE had a good ECL performance.

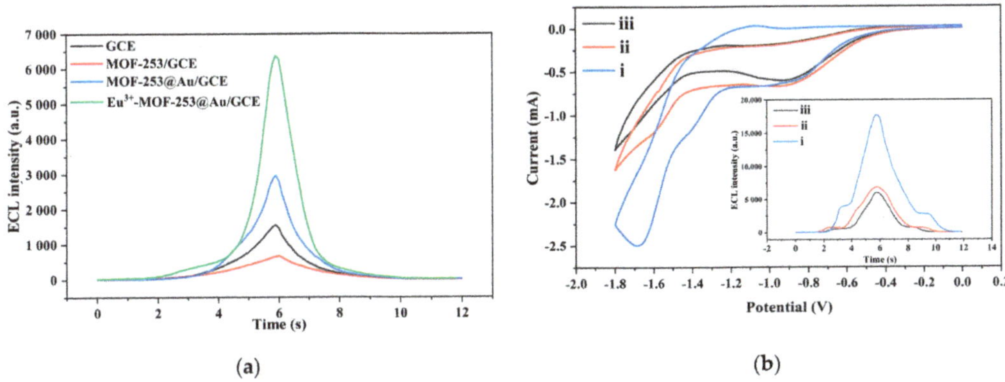

Figure 1. (**a**) ECL curves of different modified electrodes (pH 7); (**b**) CV curves of Eu^{3+}-MOF-253@Au/GCE in working solution (i), working solution with 100 μg L^{-1} carbaryl (ii), and 100 μg L^{-1} 1-naphthol (iii); the insert was the corresponding ECL curves (pH 9).

The principle of ECL detection of carbaryl was speculated as follows, according to previous literature reports: when the initial negative potential was applied to the electrode surface, the Eu^{3+} fixed on the electrode surface was charged and reduced to Eu^{2+}. Meanwhile, the co-reactant $S_2O_8{}^{2-}$ in the solution diffused to the electrode surface, and was reduced to strong oxidizing anions $SO_4{}^{-\bullet}$ and $SO_4{}^{2-}$. Subsequently, the $SO_4{}^{-\bullet}$ reacted with Eu^{2+}, and transferred energy to Eu^{2+} to generate an excited state $(Eu^{3+})^*$; then $(Eu^{3+})^*$ returned to the ground state to release photons. This process is illustrated in Equations (1)–(4), where *hv* represents the photon released when the excited state returns to the ground state. However, carbaryl was added to the working solution. and it decomposed rapidly to 1-naphthol in alkaline solution due to the electric, as well as the catalytic effect of nanogold [40,41]. The specific recognition of 1-naphthol by Eu^{3+}-MOF-253@Au was speculated to have several possibilities: first, the 1-naphthol obtained from the hydrolysis of carbaryl was more active and more amenable to electrochemical reactions due to the presence of the naphthalene ring [42]; second, an electron transfer between Eu^{3+}-MOF-253@Au and 1-naphthol [43]; third, an energy transfer from the excited state of $(Eu^{3+})^*$ was transferred to 1-naphthol or the oxide it formed [44,45]; fourth, electrostatic interactions between the free Lewis base site of the pyridine ring in Eu^{3+}-MOF-253@Au and the hydroxyl group of 1-naphthol [32,46]; and fifth, the possible formation of a complex between 1-naphthol and Eu^{3+} that prevents the reduction of exposed Eu^{3+} in the material to Eu^{2+}. This leads to the bursting of the ECL. The above process can be represented by Scheme 2. Meanwhile, this conjecture was verified using cyclic voltammetry, and the result is presented in Figure 1b. Curves (i), (ii), and (iii) were the CV curves of the Eu^{3+}-MOF-253@Au/GCE in a working solution with 100 μg L^{-1} carbaryl, and with 100 μg L^{-1} 1-naphthol, respectively. It can be seen that the original obvious redox peak became significantly smaller after the addition of the target, probably because the hydroxyl group of naphthol complexed with Eu^{3+}, which blocked the pathway of Eu^{3+} participation in the ECL process.

$$Eu^{3+} + e^{-} \rightarrow Eu^{2+} \quad (1)$$

$$S_2O_8^{2-} + e^{-} \rightarrow SO_4^{-\bullet} + SO_4^{2-} \quad (2)$$

$$Eu^{2+} + SO_4^{-\bullet} \rightarrow (Eu^{3+})^* + SO_4^{2-} \quad (3)$$

$$(Eu^{3+})^* \rightarrow Eu^{3+} + h\nu \quad (4)$$

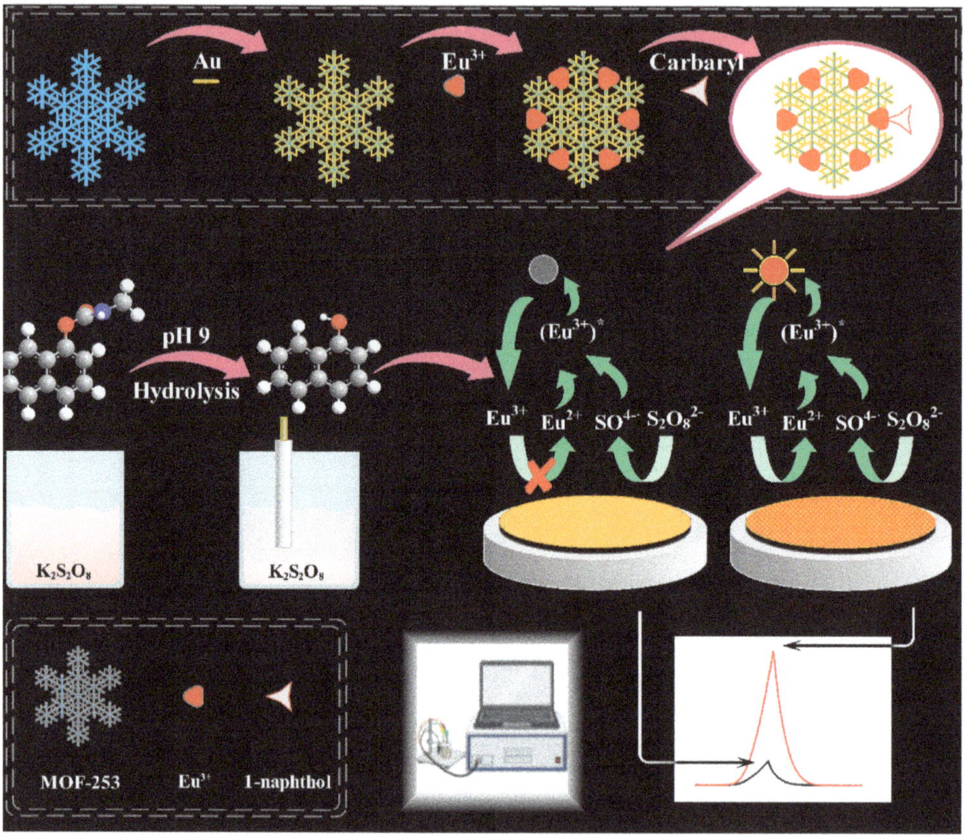

Scheme 2. Schematic diagram of the synthesis of Eu^{3+}-MOF-253@Au/GCE and carbaryl detection.

3.2. Optimization of ECL Detection Platform

There were various factors affecting the detection results, but the detection principle of this method was based on the rapid hydrolysis of carbaryl in solution; as such, the selection of the pH value of the detection solution was crucial. Additionally, since pH also affects the strength of the ECL signal, a high pH might favor the formation of the $(Eu^{3+})^*$ [33]. Taking the above into consideration, the pH of the detection solution was adjusted for detection from 7, gradually increasing in order to achieve the optimal level of ECL signal for the constructed sensor, taking into account the practical operation, the electrode working environment, and the stability of the synthesized material. The results are shown in Figure 2a. The ECL signal gradually increased during the increase in pH from 7 to 9, and the intensity of the ECL signal had reached an appreciable 17,825 a.u. when the pH reached 9. The corresponding CV curves in Figure 2a show that the peak current increased with increasing pH, which may be because the increase in pH contributed to the redox reaction in the system, which, in turn, increased the electron transfer and generated more excited states $(Eu^{3+})^*$, thereby increasing the ECL signal intensity. At the same time, the ECL quenching values of carbaryl and 1-naphthol to the sensor were compared and analyzed at different pH values. The results showed that when the pH value reached 9, the quenching values of carbaryl were consistent with that of 1-naphthol at the corresponding concentration Therefore, the pH of the working solution was adjusted to 9 in the subsequent measurements.

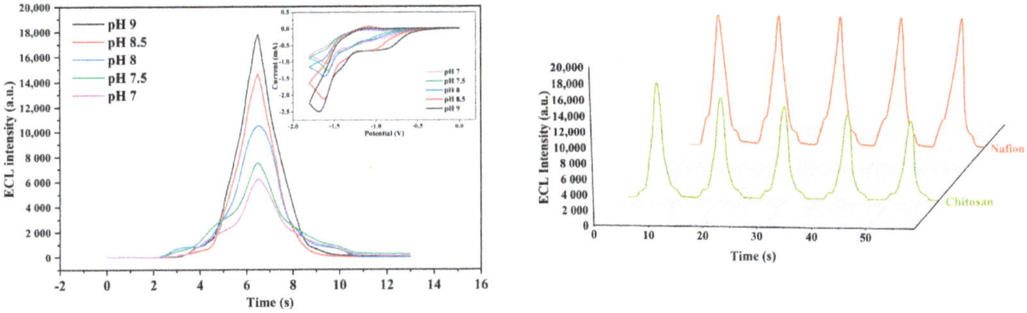

(a) (b)

Figure 2. (**a**) Optimization of pH value of working solution, insert was CV curves at different pH values; (**b**) ECL curves of electrodes modified with Nafion and chitosan, respectively.

In order to immobilize the synthesized material on the electrode surface without affecting the detection of the electrode, both Nafion and chitosan solutions (1%), as common electrode modifiers in electrochemical detection, were used to immobilize Eu^{3+}-MOF-253@Au onto the GCE surface to prepare the corresponding electrodes. As shown in Figure 2b, the ECL signal of the chitosan-modified electrode was weaker, and gradually decreased compared with that of the Nafion-fixed electrode, probably because the material was shed during the detection process, indicating that chitosan was not suitable for the fixation of the electrode modification material prepared in this study. On the contrary, the ECL signal of the electrode modified by the Nafion solution was more stable and remained above 17,000 a.u; therefore, the Nafion solution was used for the immobilization of the electrode material in the subsequent experiments.

3.3. *Performance of ECL Detection Platform*

The electrode was modified and assayed according to the conditions obtained from the above optimization experiments to evaluate its recognition, as well as the detection ability for the targets. Figure 3a shows that the ECL signal intensity of the prepared electrode gradually decreased with increasing carbaryl concentration in the range of 0.2–200 μg L^{-1}, and the logarithm of the carbaryl concentration (mg L^{-1}) showed a good linear relationship with the relative change value of ECL signal intensity $(F_0 - F)/F_0$ of the Eu^{3+}-MOF-253@Au/GCE. Where F_0 represents the initial fluorescence intensity of the Eu^{3+}-MOF-253@Au/GCE in the working solution, and F represents the fluorescence intensity obtained by the Eu^{3+}-MOF-253@Au/GCE detecting a certain concentration of the target solution. The linear regression equation was as follows: $(F_0 - F)/F_0 = 0.21956\lg C + 0.84778$ (R^2 = 0.9988), and the detection limit was calculated as 0.14 μg L^{-1}.

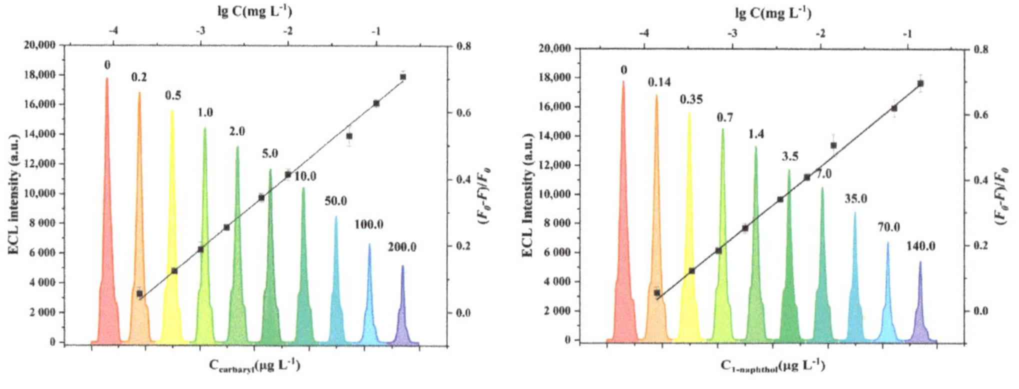

Figure 3. (**a**) ECL curves of a series concentration of carbaryl (0.2, 0.5, 1.0, 2.0, 5.0, 10.0, 50.0, 100.0, and 200.0 μg L^{-1}) detected by Eu^{3+}-MOF-253@Au sensor, and the calibration curve for carbaryl detection; (**b**) ECL curves of a series concentration of 1-naphthol (0.14, 0.35, 0.7, 1.4, 3.5, 7.0, 35.0, 70.0, and 140.0 μg L^{-1}) detected by Eu^{3+}-MOF-253@Au sensor, and the calibration curve for 1-naphthol detection.

In order to verify whether the ECL signal variation was due to 1-naphthol obtained by the hydrolysis of carbaryl, 1-naphthol solutions with corresponding concentrations were prepared and measured under the same conditions (based on the molar mass calculation, hydrolysis of 1 mg carbaryl yields around 0.7165 mg of 1-naphthol; therefore, the solutions of 0.07, 0.14, 0.35, 0.7, 1.4, 3.5, 7.0, 35.0, and 70.0 μg L^{-1} of 1-naphthol were configured). The curves in Figure 3b show the variation of the ECL signal intensity when the electrode measured 0.14–140 μg L^{-1} of 1-naphthol solution, as well as the calibration curve plotted from the signal variation calculation. The regression curve equation for 1-naphthol was calculated as follows: $(F_0 - F)/F_0 = 0.21996\lg C + 0.87963$ (R^2 = 0.9983). The change of the ECL curve in Figure 3b essentially matches the trend of the change of the ECL curve in Figure 3a, and the regression equation of the fitted curve in Figure 3b also almost agrees with the slope of the equation obtained from the detection of carbaryl, which means that it can essentially be concluded that 1-naphthol can lead to the change of the ECL signal intensity of the Eu^{3+}-MOF-253@Au/GCE.

3.4. *Selectivity, Precision and Stability*

The ability to recognize the target is one of the important properties of the sensor, so several carbamate pesticides (metolcarb, propoxur, isoprocarb, fenobucarb, methomyl) were selected as interferents to evaluate the selectivity of the prepared electrodes. The interferents' chemical structure formulae are shown in Figure 4a. Figure 4b shows that the ECL quenching value ($F_0 - F$) of the prepared Eu^{3+}-MOF-253@Au/GCE for carbaryl (1 μg L^{-1}) was several times higher than that for other analytes of the same concentration. The ECL quenching values ($F_0 - F$) of the sensor increased more in the presence of interferents than in the absence of interferents, which may be due to the effect of the similar structure of the interferents. The quenched values ($F_0 - F$) with and without the interferents were analyzed by analysis of variance (ANOVA), and the *F*-test showed no significant difference in the values of ECL signal changes in the presence and absence of the interferents. It was probably because they and their hydrolysis products were not able to complex with Eu^{3+} chelated on MOF-253, indicating the good selectivity and anti-interference performance of the prepared sensor.

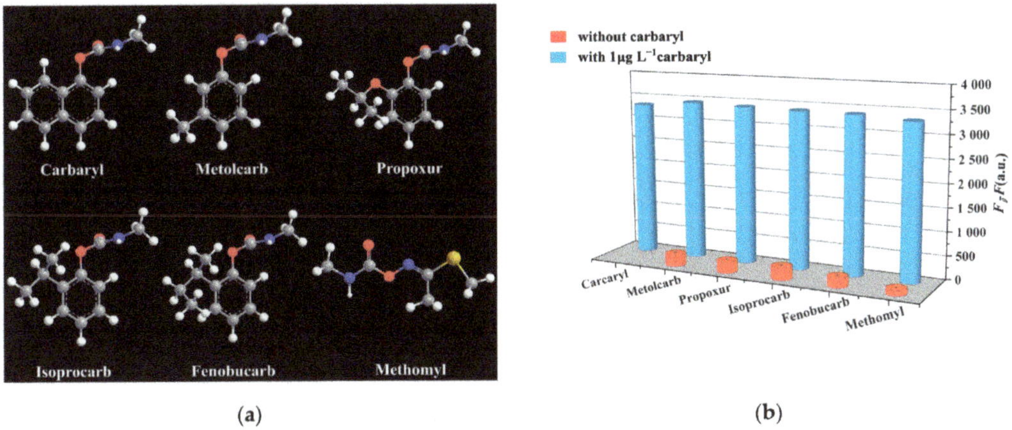

Figure 4. (**a**) The chemical structure of carbaryl and its structural analogs; (**b**) the selectivity of the Eu^{3+}-MOF-253@Au/GCE towards carbaryl and interferents.

To evaluate the precision of the constructed sensors, six electrodes were prepared using the same modification method for parallel experiments. The initial signal intensity F_0, the ECL signal intensity F for 1 μg L^{-1} carbaryl, and the difference ($F_0 - F$) were determined and analyzed separately. It can be seen from Figure 5a that the prepared Eu^{3+}-MOF-253@Au sensor has a stable ECL signal. The ANOVA performed on the ($F_0 - F$) values revealed that there was no significant difference in the detection of carbaryl by the six electrodes, and that the prepared sensor has reliable reproducibility (The F-test results were: $F = 0.4932 < 3.106 = F(6\text{–}1, 18\text{–}6)$), $\alpha = 0.05$.

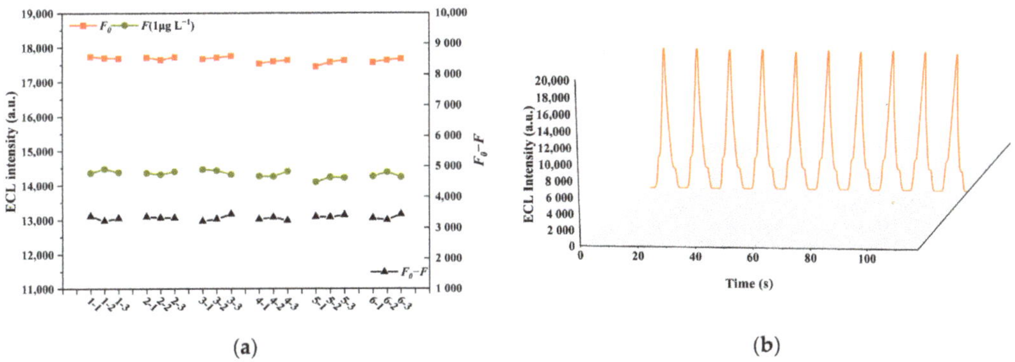

Figure 5. (**a**) The precision of the Eu^{3+}-Au@MOF-253 sensor; (**b**) the ECL curve for 10 consecutive detections carried out by Eu^{3+}-MOF-253@Au/GCE.

Stability is also an important index of electrochemical sensors, so the same electrode was measured 10 times continuously, and its signal change was analyzed to assess the stability of the Eu^{3+}-MOF-253@Au sensor. In addition, the prepared sensor, after measuring its ECL signal, was stored at 4 °C, and its signal intensity was measured every 6 days to evaluate the stability after long-term storage in three groups operated in parallel. The results are shown in Figure 5b and Figure S3, respectively. The Eu^{3+}-MOF-253@Au sensor displayed excellent stability for 10 consecutive measurements (Relative Standard Deviation, RSD = 0.65%) and long-term storage (RSD = 1.32%). The low RSD indicated a good reproducibility and high precision of the sensor. Moreover, a T-test was performed on the same Eu^{3+}-MOF-253@Au/GCE for inter-day and intra-day measurements of carbaryl at 1 μg L^{-1}. The results showed no significant difference between the two datasets. This result again demonstrates the accuracy of the established assay.

3.5. Real Sample Analysis

According to the regulations, the EU's maximum residue levels (MRL) of carbaryl residue in milk is 0.05 mg kg^{-1} [47], which is the same as China's, while the EU's carbaryl residue in soybeans and China's carbaryl residue in soybean oil is 0.05 mg kg^{-1} and 1 mg kg^{-1}, respectively [48]. Milk and soybean oil were used as actual samples for spike recovery experiments, and the ECL determination results were verified by HPLC simultaneously. The results of the analysis and calculation are presented in Table 1, and the recoveries were calculated to be 76.5–95.4% for three different concentration levels spiked on two actual samples. A paired sample T-test between spiked and detected concentrations showed $p = 0.07174 > 0.05$, indicating that there was no significant difference between the spiked and detected concentrations. This result proved that the constructed Eu^{3+}-Au@MOF-253 sensor can be applied to the analysis of real samples.

Table 1. Determination of carbaryl in food samples by the proposed ECL and HPLC.

Sample	Added (μg kg^{-1} /μg L^{-1})	HPLC		ECL	
		Found (μg kg^{-1} /μg L^{-1})	Recovery (Mean ± SD, n = 3)	Found (μg kg^{-1} /μg L^{-1})	Recovery (Mean ± SD, n = 3)
Milk	0	not detected	—	not detected	—
	1	0.87	87.1 ± 1.5	0.82	82.4 ± 2.4
	50	47.42	94.8 ± 3.4	47.57	95.1 ± 3.6
	100	97.73	97.7 ± 1.9	95.42	95.4 ± 2.2
Soybean oil	0	not detected	—	not detected	—
	1	0.83	82.7 ± 2.6	0.77	76.5 ± 2.5
	50	44.12	89.1 ± 2.6	41.53	83.1 ± 3.1
	100	93.51	93.5 ± 1.8	86.33	86.3 ± 1.7

n: the number of measurements.

3.6. Comparison of Different Detection Methods

The performance of the trace carbaryl detection strategy based on the constructed Eu^{3+}-MOF-253@Au sensor was compared with previously reported carbaryl detection methods, the result of which is presented in Table 2. It is clear that the established ECL method can balance high detection efficiency, large detection range, and low detection limit with excellent performance. It also showed lower limit of detection (LOD), as well as good recoveries and low RSDs in food samples compared to other ECL methods that selected water as the actual sample. This was attributed to the following aspects: (1) the excellent thermal and chemical stability of the MOFs material laid the structural foundation for the construction, modification, and long-term storage of the sensor; (2) the excellent chemical and fluorescence properties of Eu^{3+} facilitated the recognition of 1-naphthol by Eu^{3+}-MOF-253, thereby avoiding the superposition of multiple recognition elements and providing good selectivity; (3) the nanogold dispersed on the backbone of the MOFs had good electrical conductivity, which improved the electrode surface electron transfer rate, and effectively amplified the ECL response; (4) placing the target carbaryl in the detection solution adjusted to alkaline was beneficial for the rapid completion of its decomposition reaction and uniform contact with the electrode, which also improved the detection efficiency. On the other hand, it was presumed that the alkaline detection solution was conducive to the formation of the excited state of the luminol, effectively enhancing the ECL intensity.

Table 2. Comparison of the proposed method with other reported methods in the literature.

Method	Detection Range	LOD	Sample	Recovery (%)	RSD (%)	Year	References
High performance liquid chromatography-diode array	1.0 to 320.0 ng mL^{-1}	0.12–0.40 ng mL^{-1}	Milk, White wine	86.5–99.2	3.2–5.7	2019	[23]
	0.5 to 160.0 ng mL^{-1}	0.06–0.20 ng mL^{-1}	Juice	82.0–104.7	2.8–4.5		
Electrospray ionization mass spectrometry	20.00–1000.00 ng mL^{-1}	1.16–4.18 ng g^{-1}	Honey	87.00–114.98	0.97–6.22	2017	[24]
DNA aptamer	100–1500 nmol L^{-1}	15.23 nmol L^{-1}	Tap water, Xuanwu Hu	97.7–107.3%	0.8–2.5	2021	[26]
Surface-enhanced Raman scattering	$10^{-4} - 10^{-9}$ g mL^{-1} (situ detection)	0.77 ng mL^{-1} (situ detection)	Cherry	–	–	2021	[27]
	$10^{-3} - 10^{-8}$ g mL^{-1} (swabbing detection)	5.8 ng mL^{-1} (swabbing detection)		–	–		
Colorimetric	10 µM-5 mM	10 ppm	Apple	–	–	2018	[28]
ECL	0–18 µM	1.0 nM	–	96.9–101.7	1.87–3.02	2013	[49]
ECL	1.0×10^{-12}–1.0×10^{-7} M	1.0×10^{-12} M	Tap water	94.0–95.8%	3.53–4.62	2020	[50]
ECL	5×10^{-4}–10 mg mL^{-1}	2×10^{-4} µg mL^{-1}	River water	99.0–108.0	5%<	2015	[51]
ECL	0.2–200 µg L^{-1}	0.14 µg L^{-1}	Milk, Soybean oil	76.5–95.4%	2.0–3.8	2021	This work

4. Conclusions

In this study, based on the selectivity of Eu^{3+}-MOF-253 for 1-naphthol, nanogold was introduced into the MOF framework to enhance the catalytic properties and conductivity of Eu^{3+}-MOF-253 to construct an ECL sensor for the sensitive detection of carbaryl. After a series of optimization and evaluation, it was found that the established assay showed admirable detection efficiency, sensitivity, low detection limit (LOD = 0.14 $\mu g\ L^{-1}$), and high stability, as well as demonstrating satisfactory recoveries in actual samples (76.5–95.4%), which implies good application prospects. This work not only proposes a rapid and sensitive method for the detection of carbaryl residues in foods, but also enriched the method for the determination of hazard factors in food, and expanded the application of MOFs in the field of electrochemistry.

Supplementary Materials: The following supporting information can be downloaded at: https://www.mdpi.com/article/10.3390/foods11101487/s1, Reagents and Materials, Apparatus and Instruments, Sample preparation, Characterization of MOF-253, MOF-253@Au and Eu^{3+}-MOF-253@Au, Calculation of detection limit. Figure S1: The SEM image of MOF-253 (A) and MOF-253@Au (B); Figure S2: (A) The FT-IR patterns of MOF-253 and MOF-253@Au. (B) The XRD patterns of MOF-253 and MOF-253@Au. (C) The X-ray photoelectron spectra of MOF-253, MOF-253@Au, and Eu-MOF-253@Au. (D) The N2 sorption isotherm at 77K for MOF-253 and Eu-MOF-253@Au; Figure S3: The variation for ECL responses of proposed sensor after long-term storage.

Author Contributions: Conceptualization, C.L.; methodology, C.L.; software, C.L.; validation, C.L.; formal analysis, C.L. and H.W.; investigation, X.H.; data curation, C.L. and Y.C.; writing—original draft preparation, C.L.; writing—review and editing, G.F.; funding acquisition, G.F. All authors have read and agreed to the published version of the manuscript.

Funding: This research was funded by the National Key Research and Development Program of China (project No. 2017YFC1600803).

Institutional Review Board Statement: Not applicable.

Informed Consent Statement: Not applicable.

Data Availability Statement: The data presented in this study are available on request from the corresponding author.

Acknowledgments: This work was supported by the National Key Research and Development Program of China (project No. 2017YFC1600803).

Conflicts of Interest: The authors declare no conflict of interest. The funders had no role in the design of the study; in the collection, analyses, or interpretation of data; in the writing of the manuscript, or in the decision to publish the results.

References

1. Cao, S.P.; Luo, Q.X.; Li, Y.J.; Liang, R.P.; Qiu, J.D. Gold nanoparticles decorated carbon nitride nanosheets as a coreactant regulate the conversion of the dual-potential electrochemiluminescence of $Ru(bpy)_3^{2+}$ for Hg^{2+} detection. *Chem. Commun.* **2020**, *56*, 5625–5628. [CrossRef] [PubMed]
2. Li, S.; Ma, X.; Pang, C.; Wang, M.; Luo, J. Novel chloramphenicol sensor based on aggregation-induced electrochemiluminescence and nanozyme amplification. *Biosens. Bioelectron.* **2018**, *176*, 112944. [CrossRef] [PubMed]
3. Liao, N.; Liu, J.; Chai, Y.; Yuan, R.; Zhuo, Y. DNA Structure Transition-Induced Affinity Switch for Biosensing Based on the Strong Electrochemiluminescence Platform from Organic Microcrystals. *Anal. Chem.* **2020**, *92*, 3940–3948. [CrossRef] [PubMed]
4. Xu, Z.; Guo, Z.; Dong, S. Electrogenerated chemiluminescence biosensor with alcohol dehydrogenase and tris(2,2′-bipyridyl)ruthenium (II) immobilized in sol–gel hybrid material. *Biosens. Bioelectron.* **2005**, *21*, 455–461. [CrossRef] [PubMed]
5. Liu, Z.; Qi, W.; Xu, G. ChemInform Abstract: Recent Advances in Electrochemiluminescence. *Cheminform* **2015**, *46*, 3117–3142. [CrossRef]
6. Miao, W. Electrogenerated chemiluminescence and its biorelated applications. *Chem. Rev.* **2008**, *108*, 2506–2553. [CrossRef]
7. Martin, J.E.; Hart, E.J.; Adamson, A.W.; Gafney, H.; Halpern, J. Chemiluminescence from the reaction of the hydrated electron with tris(bipyridyl)ruthenium(III). *J. Am. Chem. Soc.* **1972**, *94*, 9238–9240. [CrossRef]
8. Xu, G.; Zeng, X.; Lu, S.; Dai, H.; Gong, L.; Lin, Y.; Wang, Q.; Tong, Y.; Chen, G. Electrochemiluminescence of luminol at the titanate nanotubes modified glassy carbon electrode. *Luminescence* **2013**, *28*, 456–460. [CrossRef]

9. Wang, Y.; Lu, J.; Tang, L.; Chang, H.; Li, J. Graphene oxide amplified electrogenerated chemiluminescence of quantum dots and its selective sensing for glutathione from thiol-containing compounds. *Anal. Chem.* **2009**, *81*, 9710–9715. [CrossRef]
10. Myung, N.; Ding, Z.; Bard, A.J. Electrogenerated Chemiluminescence of CdSe Nanocrystals. *Nano Lett.* **2002**, *2*, 1315–1319. [CrossRef]
11. Yang, L.; Zhang, B.; Fu, L.; Fu, K.; Zou, G. Efficient and Monochromatic Electrochemiluminescence of Aqueous-Soluble Au Nanoclusters via Host–Guest Recognition. *Angew. Chem. Int. Ed.* **2019**, *131*, 6975–6979. [CrossRef]
12. Chen, S.; Lv, Y.; Shen, Y.; Ji, J.; Zhou, Q.; Liu, S.; Zhang, Y. Highly Sensitive and Quality Self-Testable Electrochemiluminescence Assay of DNA Methyltransferase Activity Using Multifunctional Sandwich-Assembled Carbon Nitride Nanosheets. *ACS Appl. Mater. Interfaces* **2018**, *10*, 6887. [CrossRef] [PubMed]
13. Danis, A.S.; Potts, K.P.; Perry, S.C.; Mauzeroll, J. Combined Spectroelectrochemical and Simulated Insights into the Electrogenerated Chemiluminescence Coreactant Mechanism. *Anal. Chem.* **2018**, *90*, 7377–7382. [CrossRef] [PubMed]
14. Li, H.L.; Eddaoudi, M.M.; O'Keeffe, M.; Yaghi, O.M. Design and Synthesis of an Exceptionally Stable and Highly Porous Metal-Organic Framework. *Nature* **1999**, *402*, 276–279. [CrossRef]
15. Lin, X.; Jia, J.; Hubberstey, P.; Schrder, M.; Champness, N.R. Hydrogen storage in metal–organic frameworks. *Crystengcomm* **2007**, *9*, 438–448. [CrossRef]
16. Horcajada, P.; Serre, C.; Maurin, G.; Ramsahye, N.A.; Balas, F.; Vallet-Regi, M.; Sebban, M.; Taulelle, F.; Ferey, G. Flexible Porous Metal-Organic-Frameworks for a controlled drug delivery. *J. Am. Chem. Soc.* **2008**, *130*, 6774–6780. [CrossRef]
17. Yang, Q.; Xu, Q.; Jiang, H.L. Metal–organic frameworks meet metal nanoparticles: Synergistic effect for enhanced catalysis. *Chem. Soc. Rev.* **2017**, *46*, 4774–4808. [CrossRef]
18. Cui, Y.; Chen, B.; Qian, G. Lanthanide metal-organic frameworks for luminescent sensing and light-emitting applications. *Coord. Chem. Rev.* **2014**, *273*, 76–86. [CrossRef]
19. Jin, Z.; Zhu, X.; Wang, N.; Li, Y.; Lei, J. Electroactive Metal Organic Frameworks as Emitters for Self-Enhanced Electrochemiluminescence in Aqueous Medium. *Angew. Chem.* **2020**, *132*, 10532–10536. [CrossRef]
20. Hu, G.B.; Xiong, C.Y.; Liang, W.B.; Zeng, X.S.; Xu, H.L.; Yang, Y.; Yao, L.Y.; Yuan, R.; Xiao, D.R. Highly Stable Mesoporous Luminescence-Functionalized MOF with Excellent Electrochemiluminescence Property for Ultrasensitive Immunosensor Construction. *ACS Appl. Mater. Interfaces* **2018**, *10*, 15913–15919. [CrossRef]
21. Castorina, R.; Harnly, M.; Eskenazi, B.; Barr, D.B.; Bradman, A. Carbaryl and naphthalene exposures among a pregnant Latina population living in an agricultural area. *Epidemiology* **2008**, *19*, S207–S208. [CrossRef]
22. Karami-Mohajeri, S.; Abdollahi, M. Toxic influence of organophosphate, carbamate, and organochlorine pesticides on cellular metabolism of lipids, proteins, and carbohydrates: A systematic review. *Hum. Exp. Toxicol.* **2011**, *30*, 1119–1140. [CrossRef] [PubMed]
23. Wang, X.; Meng, X.; Wu, Q.; Wang, C.; Wang, Z. Solid phase extraction of carbamate pesticides with porous organic polymer as adsorbent followed by high performance liquid chromatography-diode array detection. *J. Chromatogr. A* **2019**, *1600*, 9–16. [CrossRef] [PubMed]
24. Deng, M.; Yu, T.; Luo, H.; Zhu, T.; Huang, X.; Luo, L. Direct detection of multiple pesticides in honey by neutral desorption-extractive electrospray ionization mass spectrometry. *Int. J. Mass Spectrom.* **2017**, *422*, 111–118. [CrossRef]
25. Yang, L.; Zhou, Y.; Qu, R.; Xu, Y.; Shang, S.; Hui, N. Non-Enzymatic Electrochemical Sensors Based on Conducting Polymer Hydrogels for Ultrasensitive Carbaryl Pesticide Detection. *J. Electrochem. Soc.* **2021**, *168*, 047506. [CrossRef]
26. Liu, Y.; Yang, G.; Li, T.; Deng, Y.; He, N. Selection of a DNA aptamer for the development of fluorescent aptasensor for carbaryl detection. *Chin. Chem. Lett.* **2021**, *32*, 1957–1962. [CrossRef]
27. Ma, Y.; Chen, Y.; Tian, Y.; Gu, C.; Jiang, T. Contrastive Study of In Situ Sensing and Swabbing Detection Based on SERS-Active Gold Nanobush–PDMS Hybrid Film. *J. Agric. Food Chem.* **2021**, *69*, 1975–1983. [CrossRef]
28. Lee, M.G.; Patil, V.; Na, Y.C.; Lee, D.S.; Lim, S.H.; Yi, G.R. Highly stable, rapid colorimetric detection of carbaryl pesticides by azo coupling reaction with chemical pre-treatment. *Sens. Actuators B Chem.* **2018**, *261*, 489–496. [CrossRef]
29. Staninski, K.; Lis, S. Ultraweak emission of the Eu(III) ions in cathodic generated electrochemiluminescence. *Opt. Mater.* **2011**, *33*, 1540–1543. [CrossRef]
30. Yu, H.X.; Cui, H.; Guan, J.B. Cathodic electrochemiluminescence of acetonitrile, acetonitrile-1,10-phenanthroline and acetonitrile-ternary Eu(III) complexes at a gold electrode. *Luminescence* **2010**, *21*, 81–89. [CrossRef]
31. Zhou, H.; Zhang, Y.Y.; Liu, J.; Xu, J.J.; Chen, H.Y. Electrochemiluminescence Resonance Energy Transfer between CdS:Eu Nancrystals and Au Nanorods for Sensitive DNA Detection. *J. Phys. Chem. C* **2012**, *116*, 17773–17780. [CrossRef]
32. Zhao, J.J.; Liu, P.Y.; Dong, Z.P.; Liu, Z.L.; Wang, Y.Q. Eu(III)-organic framework as a multi-responsive photoluminescence sensor for efficient detection of 1-naphthol, Fe^{3+} and MnO_4^- in water. *Inorg. Chim. Acta* **2020**, *511*, 119843. [CrossRef]
33. Li, D.; Shan, Y.; Xu, J.J.; Chen, H.Y. Electrochemiluminescence behaviors of Eu^{3+}-doped CdS nanocrystals film in aqueous solution. *Nanoscale* **2012**, *4*, 831–836. [CrossRef]
34. Bloch, E.D.; Britt, R.; Lee, R.; Doonan, R.J.; Uribe-Romo, R.J.; Furukawa, H.; Long, R.R.; Yaghi, R.M. Metal insertion in a microporous metal-organic framework lined with 2,2′-bipyridine. *J. Am. Chem. Soc.* **2010**, *132*, 14382. [CrossRef]
35. Deng, X.; Qin, Y.; Hao, M.; Li, Z. MOF-253-Supported Ru Complex for Photocatalytic CO_2 Reduction by Coupling with Semidehydrogenation of 1,2,3,4-Tetrahydroisoquinoline (THIQ). *Inorg. Chem.* **2019**, *58*, 16574–16580. [CrossRef]
36. Shi, L.; Xiang, Z.; Liu, T.; Zhao, H.; Lan, M. Encapsulating Cu nanoparticles into metal-organic frameworks for nonenzymatic glucose sensing. *Sens. Actuators B Chem.* **2016**, *227*, 583–590. [CrossRef]

37. Si-Jia; Qin; Bing; Yan, A facile indicator box based on Eu^{3+} functionalized MOF hybrid for the determination of 1-naphthol, a biomarker for carbaryl in urine. *Sens. Actuators B Chem.* **2018**, *259*, 125–132. [CrossRef]
38. *GB 23200.90-2016*; National Food Safety Standards-Determination of Multiple Carbamate Pesticides Residues in Milk and Dairy Products Liquid Chromatography-Mass Spectrometry. National Health and Family Planning Commission of the People's Republic of China, Ministry of Agriculture of the People's Republic of China, State Administration for Market Regulation: Beijing, China, 2016; p. 2.
39. *GB 23200.112-2018*; National Food Safety Standards-Determination of 9 Carbamate Pesticides and Metabolites Residues in Foods of Plant Origin-Liquid Chromatography-Post-Colum Derivatization. National Health Commission of the People's Republic of China, Ministry of Agriculture and Rural Affairs of the People's Republic of China, State Administration for Market Regulation: Beijing, China, 2018; p. 3.
40. Shi, X.; Liu, H.; Zhang, M.; Yang, F.; Li, J.; Guo, Y.; Sun, X. Ultrasensitive electrochemiluminescence aptasensor based on AuNPs@MWCNTs and Au@AgNPs for detection of profenofos residues. *Sens. Actuators B Chem.* **2021**, *348*, 130663. [CrossRef]
41. Wang, D.; Liang, Y.; Su, Y.; Shang, Q.; Zhang, C. Sensitivity enhancement of cloth-based closed bipolar electrochemiluminescence glucose sensor via electrode decoration with chitosan/multi-walled carbon nanotubes/graphene quantum dots-gold nanoparticles. *Biosens. Bioelectron.* **2019**, *130*, 55–64. [CrossRef]
42. Chen, W.; Liu, Y.; Zhang, Y.; Fang, J.; Xu, P.; Xu, J.; Li, X.; Liud, C.-C.; Wen, W. Highly effective and specific way for the trace analysis of carbaryl insecticides based on Au42Rh58 alloy nanocrystals. *J. Mater. Chem. A* **2017**, *5*, 7064–7071. [CrossRef]
43. Zhang, F.; Yi, W.; Chu, T.S.; Wang, Z.H.; Yang, Y.Y. A facile fabrication of electrodeposited luminescent MOF thin films for selective and recyclable sensing of nitroaromatic explosives. *Analyst* **2016**, *141*, 4502. [CrossRef] [PubMed]
44. Tao, C.L.; Chen, B.; Liu, X.G.; Zhou, L.J.; Zhu, X.L.; Cao, J.; Gu, Z.G.; Zhao, Z.; Shen, L.; Tang, B.Z. A highly luminescent entangled metal–organic framework based on pyridine-substituted tetraphenylethene for efficient pesticide detection. *Chem. Commun.* **2017**, *53*, 9975–9978. [CrossRef] [PubMed]
45. Cao, L.; Shi, F.; Zhang, W.; Zang, S.; Mak, T.C.W. Selective Sensing of Fe(3+) and Al(3+) Ions and Detection of 2,4,6-Trinitrophenol by a Water-Stable Terbium-Based Metal-Organic Framework. *Chemistry* **2015**, *21*, 15705–15712. [CrossRef] [PubMed]
46. Guo, X.Y.; Zhao, F.; Liu, J.J.; Liu, Z.; Wang, Y.Q. An ultrastable zinc(II)–organic framework as a recyclable multi-responsive luminescent sensor for Cr(III), Cr(VI) and 4-nitrophenol in the aqueous phase with high selectivity and sensitivity. *J. Mater. Chem. A* **2017**, *5*, 20035–20043. [CrossRef]
47. 1096/2014, R.E.N. EU Pesticides Database, Pesticide Residues. Available online: https://ec.europa.eu/food/plant/pesticides/eu-pesticides-database/mrls/?event=search.pr (accessed on 1 April 2021).
48. *GB2763-2021*; State Administration of Market Supervision and Administration, National Food Safety Standard-Maximum Residue Limits for Pesticides in Food. National Health Commission of the People's Republic of China, Ministry of Agriculture and Rural Affairs of the People's Republic of China, State Administration for Market Regulation: Beijing, China, 2021.
49. Luo, J.; Zhou, C.; Shi, Y.; Zhang, L.; Xiao, D. The self-assembled $Ru(bpy)_3(PF_6)_2$ nanoparticle on polystyrene microfibers and its application for ECL sensing. *Analyst* **2013**, *138*, 6171–6176. [CrossRef] [PubMed]
50. Li, L.; Zhou, L.; Liu, X.; You, T. Ultrasensitive self-enhanced electrochemiluminescence sensor based on novel PAN@Ru@PEI@Nafion nanofiber mat. *J. Mater. Chem. B* **2020**, *8*, 3590–3597. [CrossRef]
51. Wang, S.; Lv, S.; Wu, W.; Guo, Z. A Novel Electrochemiluminescence Sensor for Sensitive Determination of Carbaryl Based on Solid Phase Microextraction at NH2–Graphene–Nafion Modified Electrode. *Aust. J. Chem.* **2014**, *68*, 793. [CrossRef]

foods MDPI

Review

Aptamer-Based Fluorescent Biosensor for the Rapid and Sensitive Detection of Allergens in Food Matrices

Liping Hong [1,2], Mingfei Pan [1,2,*], Xiaoqian Xie [1,2], Kaixin Liu [1,2], Jingying Yang [1,2], Shan Wang [1,2] and Shuo Wang [1,2]

1 State Key Laboratory of Food Nutrition and Safety, Tianjin University of Science and Technology, Tianjin 300457, China; honglpstu@163.com (L.H.); qianxx8135@163.com (X.X.); Liukx2019@163.com (K.L.); yangjy0823@126.com (J.Y.); wshan0929niu@163.com (S.W.); s.wang@tust.edu.cn (S.W.)
2 Key Laboratory of Food Nutrition and Safety, Ministry of Education of China, Tianjin University of Science and Technology, Tianjin 300457, China
* Correspondence: panmf2012@tust.edu.cn; Tel.: +86-22-6091-2493

Abstract: Food allergies have seriously affected the life quality of some people and even endangered their lives. At present, there is still no effective cure for food allergies. Avoiding the intake of allergenic food is still the most effective way to prevent allergic diseases. Therefore, it is necessary to develop rapid, accurate, sensitive, and reliable analysis methods to detect food allergens from different sources. Aptamers are oligonucleotide sequences that can bind to a variety of targets with high specificity and selectivity, and they are often combined with different transduction technologies, thereby constructing various types of aptamer sensors. In recent years, with the development of technology and the application of new materials, the sensitivity, portability, and cost of fluorescence sensing technology have been greatly improved. Therefore, aptamer-based fluorescence sensing technology has been widely developed and applied in the specific recognition of food allergens. In this paper, the classification of major allergens and their characteristics in animal and plant foods were comprehensively reviewed, and the preparation principles and practical applications of aptamer-based fluorescence biosensors are summarized. In addition, we hope that this article can provide some strategies for the rapid and sensitive detection of allergens in food matrices.

Keywords: allergen; detection; aptamer; fluorescence; food safety

Citation: Hong, L.; Pan, M.; Xie, X.; Liu, K.; Yang, J.; Wang, S.; Wang, S. Aptamer-Based Fluorescent Biosensor for the Rapid and Sensitive Detection of Allergens in Food Matrices. *Foods* **2021**, *10*, 2598. https://doi.org/10.3390/foods10112598

Academic Editor: Pratik Banerjee

Received: 30 September 2021
Accepted: 22 October 2021
Published: 27 October 2021

1. Introduction

Food allergies, an adverse reaction to antigenic substances in food mediated by the immune system, have been recognized as a global health issue with increasing prevalence in the field of food safety [1,2]. Most food allergies are immunoglobulin (Ig) E-mediated type I (immediate type) hypersensitivity reactions [3]. An epidemiological survey by the institute of infectious diseases shows that about 6–9.3% of children and 3.4–5.0% of adults have food allergies, which means the incidence of food allergies in infants and children is generally higher than that of adults [4–6]. However, there is still no standard cure for food allergies except avoiding eating foods that contain allergens. Therefore, the development of rapid and effective detection methods for allergens in food matrices is a topic of concern in the whole society.

In the past few decades, many mature techniques have been widely used in the detection of food allergens, such as the enzyme-linked immunosorbent assay (ELISA), liquid chromatography-mass spectrometry (LC-MS), and polymerase chain reaction (PCR) [7–9]. The ELISA method has been widely used in the detection of food allergens due to its high specificity and sensitivity. Nevertheless, due to the influence of various external conditions such as food processing methods, there would be false positive and false negative results [10,11]. Moreover, PCR method is usually used for monitoring allergic components in food processing due to its high specificity and high automation. However,

PCR technology is not suitable for identifying allergen proteins with unascertained genes, which limits its scope of application [12,13]. Furthermore, HPLC and LC-MS are standard strategies for the quantitative analysis of allergens in various food matrices. Because of the precision requirements of the instruments, these methods usually require strict sample pre-treatment processes, a larger sample volume, and a longer analysis time, resulting in a higher detection cost [14]. Currently, biosensors with high sensitivity and specificity, such as surface-enhanced Raman spectroscopy (SERS), electrochemical biosensors, and quartz crystal microbalance (QCM) biosensors, can rapidly analyze and screen food allergens and allow on-site analysis, which are considered effective detection technology [15–17]. However, these biosensors usually require expensive instruments, proficient operators, and higher requirements for the surrounding environment. Therefore, there is an urgent need to develop rapid, accurate, sensitive, and easy-to-operate detection methods to quantify allergens in food matrices.

Nucleic acid aptamer is a nucleic acid sequence that can specifically recognize the target, screened by systematic evolution of ligands by exponential enrichment (SELEX) in vitro [18]. The combination of aptamer and target is achieved through single-stranded oligonucleotide deoxyribonucleic acid (DNA) or ribonucleic acid (RNA) folded into a specific three-dimensional structure (stem-loop, hairpin and G-quadruplex and other spatial conformations) [19–21]. Regardless of the technical requirements for the preparation of aptamers, the convenience and timeliness far exceed those of antibodies. Moreover, the screened aptamers can be artificially synthesized, which is easy to achieve standardization. In recent years, aptamers have received extensive attention due to their veracity, high specificity, and affinity, and they have been used in disease diagnosis and treatment, drug delivery, food safety testing, and environmental monitoring [22–24]. In terms of food safety, the application of aptamers to the detection of allergens in food matrices is expected to achieve the goal of accurate, rapid, and low-cost detection of allergens.

Fluorescence detection technology, due to its low cost, high sensitivity, simple performance, has attracted wide attention [25–27]. Combining fluorescence detection technology with aptamers, the development of biosensors with high sensitivity and simple detection procedures provides a feasible strategy for the detection of food allergens [28]. Aptamer-based fluorescence sensing detection is a relatively common analysis method. The fluorophore is combined with the aptamer in a labeled or non-labeled manner, and the analyte concentration and other information are reflected by the interaction of the excitation light and the identification element [29,30]. Furthermore, fluorescence intensity, decay rate, spectral properties, and fluorescence anisotropy can be used alone or in combination as signal detection means. Therefore, the method has the advantages of rapid, accurate, multi-targets, high sensitivity, simple instrumentation, and in-situ detection [31,32]. This review introduces the allergens in different food matrices in detail, explains the preparation principle of aptamer fluorescence sensing technology and its application in the detection of allergens in food matrices, and looks forward to the future development of food allergen detection methods.

2. Classification of Food Allergens

A great variety of food allergens exist widely in nature. According to the source of food allergens, they can be classified into animal allergens, plant allergens, and fungal allergens. Since the occurrence of fungal allergens is not very common, only animal and plant allergens are discussed in this review. Table 1 lists the classification of major food allergens, allergy symptoms, and other information. Specific information about food allergies is also discussed in the following sections.

Table 1. The major allergens in food matrices and their allergenic properties.

Food	Major Allergens	Molecular Mass (kDa)	Types of Proteins	The Structure of Proteins	Allergy Symptoms	Reference
Fish	Pan h 1	10–13	Calbindin	Contains 3 EF-hand regions (a motif composed of a 12-residue loop with a 12-residue-α-helix domain on each side), 2 of which can bind calcium.	Blushing, hives, nausea, stomach pain, and intestinal bleeding.	[33]
Shellfish	Cra c 1	33–39	Protein bound to actin	Adopting an α-helix structure, two molecules are entangled with each other to form a parallel dimeric α-helix structure.	Nausea, diarrhea, abdominal pain, and muscle paralysis.	[34]
	Cra c 2	38–45	Phosphoglycoprotein	Arginine kinase consists of an N-terminal domain (1–111) and a C-terminal domain (112–357). The N-terminal domain is all α-helices, and the C-terminal domain is an 8-strand anti-parallel β-sheet structure surrounded by 7 α-helices.		[35]
Milk	Bos d 8	57–37.5	Phosphate calcium binding protein	Consists of 4 independent proteins: αs1-casein, αs2-casein, β-casein, and κ-casein.	Skin rash, urticaria, eczema, vomiting, diarrhea, abdominal cramps, etc.	[36]
	Bos d 4	14.4	Combine with metal ions and participate in lactose synthesis	With a two-piece structure containing α-single loop and 310 helix larger subdomain.		[37]
	Bos d 5	18	Lipid transporter	Consists of two subunits connected by non-covalent bonds, mainly in the form of dimers.		[38]
Egg	Gal d1	28	Phosphoglycoprotein	Contains 3 independent homologous structural energy domains, and 3 functional domains are arranged consecutively in space.	Eczema, dermatitis, urticaria, vomiting, diarrhea, gas-troesophageal reflux, etc.	[39]
	Gal d2	45	Phosphoglycoprotein	Containing 4 free sulfhydryl groups, composed of 385 amino acid residues, these amino acid residues are twisted and folded to form a spherical structure with high secondary structure, most of which are α-helix and β-sheet.		[40]
	Gal d3	77	Iron-binding glycoprotein	Consisting of 686 amino acids, including 12 disulfide bonds, the N-terminal and C-terminal 2 domains each contain a binding site for Fe^{3+}.		[41]
	Gal d4	14.3	Basic globulin	A single peptide chain composed of 18 kinds of 129 amino acid residues, with 4 pairs of disulfide bonds to maintain the enzyme configuration, with lysine at the N-terminus and leucine at the C-terminus.		[42]

Table 1. *Cont.*

Food	Major Allergens	Molecular Mass (kDa)	Types of Proteins	The Structure of Proteins	Allergy Symptoms	Reference
Peanut	Ara h 1	63.5	7S Globulin	The secondary structure contains β-turns, and the quaternary structure is a trimeric complex formed by 3 monomers.	Angioedema, hypotension, asthma, anaphylactic shock, etc.	[43]
	Ara h 2	17–20	2S Albumin	A monomeric protein.		[44]
	Ara h 3	57	11S Globulin	The N-terminal and C-terminal domains of the monomer form contain 2 ciupin folds (composed of two sets of parallel β-turns, random coils and 3 α-helices).		[45]
Wheat	Tri a 36	40	Gluten	-	Wheat exercise stimulates allergies, urticaria, dermatitis, bread asthma, nausea, and diarrhea.	[46]
Soybean	Gly m 5	150–200	7S Globulin	Trimer composed of α′-subunit, α-subunit and β subunit.	Red and itchy skin, asthma and allergic rhinitis, abdominal pain, diarrhea, etc.	[47]
	Gly m 6	320–360	11S Globulin	A hexamer composed of the interaction of G1, G2, G3, G4, and G5 subunits.		[48]
Nuts	Ana o 1	50	7S legumin	Exist as a trimer in natural state.	Metallic taste in the mouth, edema of the tongue or throat, difficulty breathing and swallowing, urticaria all over the body, flushing of the skin, cramping abdominal pain, nausea.	[49]
	Jug r 2	44		Consists of 593 amino acid residues.		[50]
	Cor a 11	48		Consists of 401 amino acid residues, with two potential N-glycosylation sites (Asn38 and Asn254) and a leader peptide of 46 amino acids.		[51]
	Ana o 3	14	2S albumin	Composed of 5 helical structures, containing 2 subunits, connected by cysteine disulfide bonds.		[52]
	Jug r 1	15–16		Consists of 142 amino acid residues.		[53]
	Jug r 4	58.1	11S globulin	Except for the first 23 amino acid residues which are predicted as signal peptides, the remaining part has a total of 507 amino acid residues.		[54]
	Cor a 9	40		Composed of 515 amino acid residues, the sequence homology with Ara h 3 is about 45%.		[55]
	Pru du 6	350		Exist in the form of hexamers, each monomer subunit is composed of one acid chain of 40 to 42 kDa and one alkaline chain of 20 kDa.		[56]

2.1. *Common Allergens in Animal Food*

2.1.1. Seafood

Seafood refers to several different groups of edible aquatic animals, including fish, crustaceans, and molluscs [57,58]. For culinary reasons, the two invertebrate groups of crustaceans and mollusks are usually combined into shellfish [59,60]. Seafood are one of the "eight major allergens" identified by the Food and Agriculture Organization of the United Nations (FAO) in 1995 [61]. Seafood allergy is a serious global public health problem with increasing prevalence, which affects 2.5% of the global population due to the rapid increase in consumption [62,63]. The exposure pathways of seafood allergy include ingestion, contact, and even inhalation [64]. However, the main method of sensitization is through ingestion after cooking or processing [65].

Among fish allergenic proteins, parvalbumin (PV, Pan h 1) belongs to one of the Ca^{2+} binding proteins and can maintain allergenicity under severe destructive conditions (such as heat, chemical denaturation, and proteolytic enzymes). α-Lineage and β-lineage are two homologous lineages of PV, in which β-lineage is the major allergen and is present in almost all teleost fishes [66,67]. Furthermore, different types of PV in fish have highly conserved structures and cross-reactivity. Indeed, a study has shown that 95% of fish allergy sufferers are allergic to PV [68]. Except for PV, aldolase (40 kDa), enolase (47–50 kDa), vitellogenin, collagen, and tropomyosin (TM, Cra c 1) are minor allergens of fish [69]. In recent years, with the deepening of studies on fish allergic diseases, many new fish allergens have been reported. In fish muscle, six proteins with molecular weights between 25 and 57 kDa, such as triose phosphate isomerase (28 kDa) and pyruvate kinase (57 kDa), have been confirmed to have Ig E binding activity, but the sensitization of these new allergens needs further study [70,71].

At present, there are many allergens in shellfish products have been identified. TM, the first major allergen found in shellfish food, which protein spatial structure is a highly stable α-helical dimer. Ig E specific binding experiments showed that 72–98% of patients are allergic to TM [72,73]. Arginine kinase (AK, Cra c 2) is also identified as the major allergen in crustacean aquatic product muscle. There were 10–51% of shrimp allergic patients exhibit positive IgE binding to purified AK. Studies have shown that although the spatial structure of AK is more complex than that of TM, damage to the spatial structure (treated by heat and pH) can reduce its sensitization [74–76]. In addition to the two major allergens mentioned above, the allergen sarcoplasmic calcium binding protein in the muscles of crustacean aquatic products has an EF hand structure similar to that of PV, which can induce 29–50% of shrimp allergic patients to be positive for Ig E binding [77]. In recent years, some new allergens, such as myosin light chain, troponin, triose phosphate isomerase, and paramyosin, which can cause mild allergic reactions, have also been isolated and identified [78,79]. However, the sensitizing properties of these proteins remain unclear and further studies are needed.

2.1.2. Milk

Milk allergy seriously affects the life of people allergic to milk and has a relatively high incidence worldwide [80]. Relevant studies have shown that about 2–7% of infants and children suffer from milk allergies due to the ingestion of milk and milk products [81]. We need to clarify the structure and sensitization mechanism of milk allergens, thereby establishing accurate and sensitive detection and analysis methods to help people prevent the occurrence of allergic diseases. The main allergens in milk are casein (Bos d 8), β-lactoglobulin (Bos d 5) and α-lactalbumin (Bos d 4) [82,83].

Casein accounts for about 80% of milk protein and exists in milk in the form of micelle, which belongs to a large class of calcium-binding phosphoproteins [84]. It is coded by different genes located on the same chromosome and is divided into four types: αs1-casein, αs2-casein, β-casein, and κ-casein, which account for 32%, 10%, 28%, and 10%, respectively [85]. The difference in structure and content of casein in human milk and cow's milk is the main cause of casein sensitization. A study has shown that about 65% of sufferers are allergic to casein [86]. Whey protein accounts for 20% of milk protein, and the main allergenic components are α-lactalbumin and β-lactoglobulin [87]. Compared with casein, whey protein has a higher secondary and tertiary structure due to the fact that they are not phosphorylated and contain intramolecular disulfide bonds, which will make them resistant to acids or enzymes, so they can pass the intestinal mucosa smoothly, be recognized by the immune cells in the human body, and trigger an immune response.

Studies have shown that about 27.6–62.8% of allergic people are caused by α-lactalbumin, and about 82% are caused by β-lactoglobulin [88–90].

2.1.3. Egg

Eggs are the second leading cause of food allergies, accounting for 35% of infants and young children and 12% of adults [91–93]. However, the symptoms of egg allergy may gradually disappear with age [94]. At present, there are four major allergens found in egg whites, including ovomucoid (OVM, Gal d1, 11%), ovalbumin (OVA, Gal d2, 54%), ovotransferrin (OVT, Gal d3, 12%), and lysozyme (Lys, Gal d4, 3.5%), which account for nearly 80% of the total egg white protein. α-Livetin (Gal d5) and yolkglycoprotein42 (Gal d6) are the major allergens in the yolk [95,96]. Most egg allergies are caused by allergens in egg white [97].

OVM is composed of 186 amino acid residues and contains three functional domains with independent homologous structures. It is reported that the third functional domain of OVM has the strongest allergenicity [98]. The 20–25% glycosyl component contained in the structure makes OVM very stable to the thermal processing and enzymatic hydrolysis of trypsin [99]. OVA is a monomeric water-soluble protein composed of 385 amino acid residues and is the most abundant protein in egg whites [100]. At present, OVA has been widely used as a model protein to study protein structure, functional properties and food allergy animal models [101]. There are reports in the literature that OVA is a stronger allergen than other egg allergens [102]. OVT, a glycoprotein, is a monoglycated polypeptide composed of 686 amino acid residues. OVT possess many biological activities due to the N-terminal and C-terminal two domains each contain a Fe^{3+} binding site, respectively [103]. In addition, OVT is very similar to serum transferrin, which easy to form stable complexes with metal ions [104]. Lys, a weak allergen, is an alkaline globulin that consists of 129 amino acid residues in a single polypeptide [105]. At present, studies on Lys mainly focus on its antibacterial properties. Although OVM and OVA are more common allergens than Lys, the high usage rate of Lys in food and drug production makes it one of the important egg white allergens in allergy research [106].

2.2. *Common Allergens in Plant Food*

2.2.1. Peanut

Peanuts are one of the more common food allergens, which often cause severe allergic reactions. Peanut allergens have high thermal stability, acid and enzymatic resistance, and the general production method cannot remove the allergenicity [107,108]. In the actual production process, food processing often requires complex production processes, which will cause cross-contamination between foods. Therefore, it is difficult to accurately determine whether some foods contain peanut allergens [109,110]. Among the 13 allergens identified in peanuts, Ara h 1, Ara h 2, and Ara h 3 are the major allergens in peanut [111]. Ara h 1, a soluble protein, accounts for approximately 12–16% of the total peanut protein, which belongs to the pea globule protein and can cause more than 90% of allergic reactions. In the natural state, Ara h 1 has two kinds of monomers and trimers, which exist in the form of soluble proteins. For the trimeric form of Ara h 1, it is a homotrimeric glycoprotein formed by connecting three monomers by hydrophobic interaction. Most epitopes are more or less hidden in the natural Lys trimer complex, which can protect the monomer from degradation, and will lead to increased allergies. Guillon et al. determined the stability of the Ara h 1 trimer structure and described its spatial structure in detail [112]. Ara h 2 belongs to blue bean protein and is also the main peanut allergen, with the content of 5.9–9.3% of the total peanut protein, which can also cause more than 90% of allergic reactions. As there are many disulfide bonds in Ara h 2, the structure is very stable [113]. In addition, Ara h 2 contains two genetic variants, Ara h 2.01 (16.7 kDa) and Ara h 2.02 (18 kDa). Ara h 3 is 62–72% similar to glycinin and also has two allogeneic proteins, Ara h 3.01 (60 kDa) and Ara h 3.02 (37 kDa), in which the serum of more than 44% of peanut allergy patients can recognize Ara h 3.01 [114,115].

2.2.2. Wheat

As one of the three major grains and the staple foods of mankind, wheat is widely cultivated all over the world [116,117]. However, wheat is also a major plant allergen, and some people have severe allergic reactions to wheat [118–121]. Although most wheat allergies can cause mild reactions, in some cases, it can be life-threatening. The protein in wheat can be divided into water-soluble albumin, salt-soluble globulin, and gluten soluble in ethanol or acid according to solubility, and the allergens in wheat mostly come from glutelin [122,123]. At present, most of the wheat allergens that have been identified belong to the α-amylase/trypsin inhibitor family, and there are 18 species: Tri a 12, Tri a 14, Tri a 17–21, Tri a 25–28, Tri a 36–37, and Tri a 41–45. Among them, Tri a 14, Tri a 19, and Tri a 36 are the major allergens [124–126].

2.2.3. Soybean

Soybean, a major plant protein source, which is extensively used in the food processing industry. Simultaneously, soybean is also one of the "eight major allergens" that identified by the FAO [127]. Studies have shown that with the increase in the use of soybean and soybean products, about 1–6% of infants and children are allergic to soybean and the incidence of soybean allergies in adults is also increasing [128,129]. There are 11 kinds of allergenic proteins found in soybeans, namely: β-conglycinin (Gly m 5), hydrophobic protein, defensive protein, inhibitory protein, SAM22, 7S globulin, glycinin (Gly m 6), 2S albumin, lectin, lipoxidase, and trypsin inhibitor [130–132]. Among them, Gly m Bd 28K and Gly m Bd 30K of the 7S globulin and the α-subunit Gly m Bd 60K of β-conglycinin are the main allergens in soybeans.

Gly m Bd 30K (34 kDa) is a water-insoluble monomolecular glycoprotein consisting of 257 amino acid residues, which can be combined with subunits of β-conglycinin through disulfide bonds to participate in the folding of soy protein [133]. Gly m Bd 28K (26 kDa) is a broad bean globulin-like protein belonging to the Cupin superfamily and contains 220 amino acid residues [134]. β-conglycinin contains three subunits, namely α- subunit (68 Ku), α'-subunit (71 Ku), and β-subunit (50 Ku), and they exist in the form of homologous or heterologous trimers. Among them, Gly m Bd 60K is very stable, which is formed by the combination of polysaccharide and aspartic acid at the N-terminus of protein [135].

2.2.4. Nuts

Nuts mainly include almonds, cashews, walnuts, hazelnuts, pistachios, Brazil nuts, etc., which can cause allergic reactions [136]. Generally used as seeds or fruits, most nut proteins belong to three conservative seed storage proteins, including 2S albumin, 7S legumin and 11S legumin. 2S albumin belongs to the group of prolamins, which have the characteristics of low molecular weight and multiple cysteine residues in their sequence. Furthermore, most allergens in the prolamin group are highly resistant to heat, pH, and gastrointestinal enzymes due to their small and compact structure [137,138]. The 7S legumin is usually a trimeric protein, which has weak stability due to lack of disulfide bonds. The 11S legumin is a mature trimeric protein, and both subunits are connected by disulfide bonds. Moreover, both 7S legumin and 11S legumin belong to the Cupin family [139].

3. Application of Aptamer-Based Fluorescence Biosensors in the Detection of Different Food Allergens

3.1. Nucleic Acid Aptamer Screening Procedure

Aptamers are obtained through in vitro screening from random DNA or RNA libraries. The screening procedure of aptamers mainly includes the steps of library establishment, incubation, isolation, amplification, single-strand preparation, and purification [140,141]. After several rounds of repeated screening, aptamers with high affinity and high specificity can be obtained. The specific procedure is shown in Figure 1 [142]. (1) To construct a chemically synthesized oligonucleotide library, each oligonucleotide molecule usually contains about 80 nucleotides, including random sequences and constant primers. (2) The constructed library is incubated with the target, and part of the target is bound to the oligonucleotide. (3) Elution and separation of unbound or weakly bound oligonucleotides determines the screening efficiency of SELEX. (4) PCR or real-time PCR is used to amplify the combined sequence. (5) The amplified sequence obtained from the reaction is used to prepare the next round of secondary library.

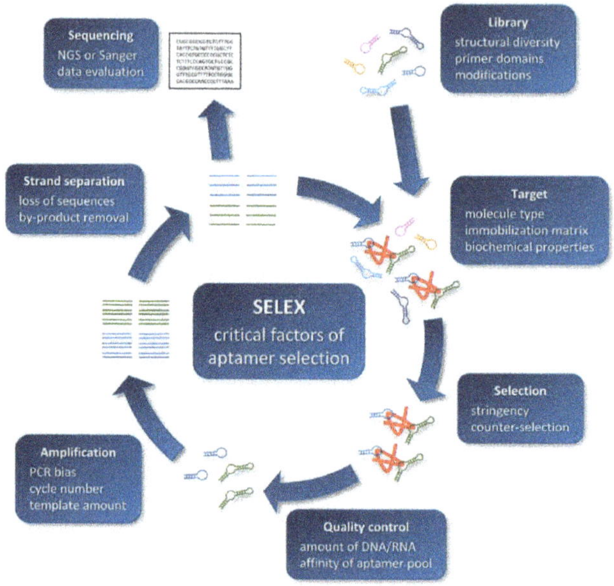

Figure 1. Schematic overview of the SELEX procedure. Reproduced with permission from [142]. Copyright Biotechnology and Applied Biochemistry, 2021.

3.2. Aptamer-Target Interaction Mechanisms

The interaction between an aptamer and its target is the core of the SELEX process and practical application. The nature of this interaction depends on the type and size of the target [143]. When the target is a small molecule, the aptamer can integrate the target into its structure through stacking (with flat, aromatic ligands and ions), electrostatic complementation (with oligosaccharides and charged amino acids), and/or hydrogen bonding interactions [144]. This interaction with small molecules gives the aptamer higher specificity that can distinguish two molecules that differ by only one methyl group, which may be due to steric hindrance. When the target is a protein macromolecule, contrary to the above situation, the aptamer will be integrated into the structure of its target or attached to the surface. Since proteins often exhibit a high degree of structural complexity, the interaction mechanism between aptamers and proteins is more diverse than that of small molecules [145,146]. In addition to hydrogen bond interactions, polar interactions and structural complementarity are also included [147]. Among them, RNA- or DNA-binding motifs that exhibit this structural complementarity are often found in nature, including helical motifs, leucine zippers, homologous domains, and beta-sheet motifs.

In addition, the nature of the interaction depends not only on the type and size of the target, but also on the structural complexity of the aptamer itself. According to the difference of sequence, aptamers may assume a polymeric state through the formation of G-tetrachromes or i-motifs, both of which can lead to interactions between multiple oligonucleotides [148]. Interestingly, according to the principle of "induced fit", the formation of aptam-target complex may involve conformational changes of the target, the aptamer, or both [149]. This principle leads to better shape complementarity, which in turn promotes stronger hydrogen bonding and van der Waals forces. Furthermore, the charge on the surface of the target also affect the interaction between the aptamer and its target. The negative charges will weaken or even prevent the binding with the aptamer, because they will adversely interact with the negatively charged phosphate groups contained in DNA and RNA. On the other hand, a positive charge can enhance the strength of the interaction, but it may also aggravate the occurrence of non-specific binding [150–152].

3.3. Detection of Animal Food Allergens

Seafood allergy is not only an important public health issue, but a serious food safety issue that affects the quality of life and may even be life threatening [64]. For people with seafood allergies, avoiding foods containing seafood allergens is still the best option. Therefore, the monitoring of allergens is a process that requires strict supervision [153]. In order to evaluate seafood allergens, new detection methods with high sensitivity and high efficiency are required.

As we all know, magnetic separation is easy to operate and can effectively reduce or eliminate the interference from complex matrices in food. Therefore, based on functionalized magnetic nanoparticles (MNPs) as a separation carrier, Zhang et al. developed a simple and versatile label-free aptamer-based fluorescent sensor for the sensitive detection of TM (Figure 2a) [154]. In the study, OliGreen dye was selected as a fluorescent signal probe. The aptamer hybridizes with the capture probe bound to the surface of the MNPs to form an aptamer-MNPs complex as detection probe. When interacting with the target, the conformation of the complex changes, resulting in the release of the aptamer from the surface of the MNPs. So, the released aptamer in the supernatant produced a significant fluorescence enhancement signal, which is because the combination of OliGreen dye and ssDNA will produce ultrasensitive and specific fluorescence enhancement phenomenon. It is worth noting that when the commercially available OliGreen dye is in the free state, the fluorescence is weak or no fluorescence, but the fluorescence will increase by more than 1000 times once combined with the aptamer ssDNA. Under the optimal conditions, the linear range was 0.4–5 $\mu g\ mL^{-1}$ (R^2 = 0.996), with a limit of detection LOD of 77 ng mL^{-1}. In addition, the highly selective aptamer-based fluorescent sensor was successfully applied to the detection of TM in food matrix. Wu et al. also developed a similar sensor with a LOD of 4.2 nM and the concentration linear from 0.5–50 $\mu g\ mL^{-1}$ [155]. Recently, Chinappan et al. developed an aptamer-based fluorescent-labeled sensor for the detection of TM. (Figure 2b) [156]. Graphene oxide (GO) is used as a platform for screening the minimum length of aptamer sequences that can bind to the target with high affinity. A fluorescein dye labeled GO quenches the truncated aptamer by π-stacking and hydrophobic interactions. After the addition of TM, the fluorescence was restored due to the competitive binding of the aptamer to GO. More importantly, the aptamer selected in this study is a truncated ligand fragment, which has four times higher affinity than the full-sequence aptamer, with a LOD of 2.5 nM. The developed aptamer-based fluorescence sensor can complete the detection within 30 min. The performance of the sensor was confirmed in the addition experiment of chicken broth, and a high percentage recovery rate (~97 $\pm$ 10%) was achieved. Compared with the above studies, the sensitivity and specificity of this work have been greatly improved.

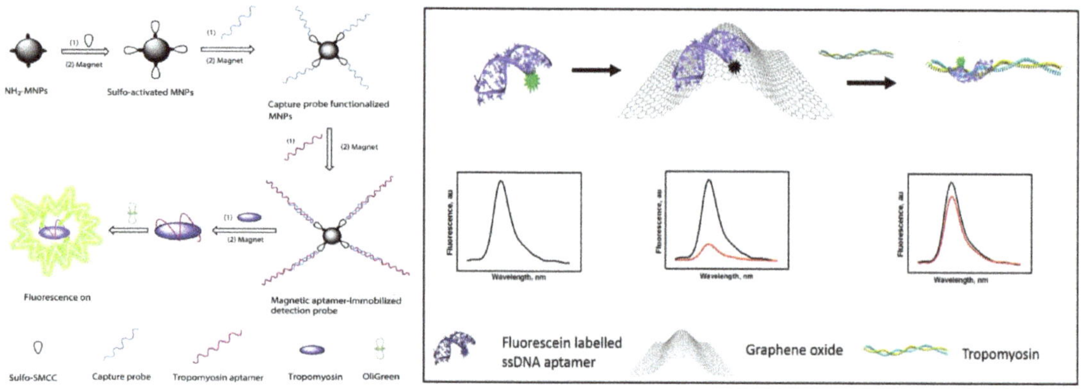

(a) (b)

Figure 2. (**a**) Schematic of preparation of magnetic-assisted fluorescent aptamer for tropomyosin detection. Reproduced with permission from [154]. Copyright Sensors and Actuators B-Chemical, 2018. (**b**) Schematic of graphene oxide-based fluorescent aptamer biosensor for TM detection A: Changes in the fluorescence intensity of the aptamer released from the GO surface; B: The linear correlation of the fluorescence intensity of TMT2 (at 515 nm) with the concentration of TM. Reproduced with permission from [156]. Copyright Food Chemistry, 2020.

Fluorescence resonance energy transfer (FRET) is a mechanism widely used in the preparation of biosensors, which is an energy transfer phenomenon between two fluorescent molecules that are very close [157]. Zhou et al. designed an aptamer-based "on-off-on" fluorescent biosensor based on FRET and used developed carboxyl functionalized carbon quantum dots (cCQDs) and GO for the detection of shellfish allergen arginine kinase (AK) (Figure 3A) [158]. The cCQDs-aptamer probe and GO self-assemble for the first time through a specific π-π interaction, so that the fluorescence of cCQDs is effectively quenched. After the addition of AK, cCQDs-aptamer is released from the GO surface and then forms the cCQDs-aptamers-AK complex, which restores the fluorescence of cCQDs. The aptamer-based FRET sensor can perform sensitive detection in the AK concentration range of 0.001–10 $\mu g\ mL^{-1}$, with a LOD of 0.14 ng mL^{-1} (S/N = 3) and a limit of quantification (LOQ) of 0.27 ng mL^{-1} (S/N = 10). Furthermore, in a control experiment with a blank sample, it was found that the sensor has high specificity. This reliable, precise, highly specific, and easy-to-operate aptamer sensor may provide a new perspective for the application of fluorescence sensing technology in the field of food safety.

In recent years, biosensors based on dual signals or functions have received widespread attention due to the diversity of detection. Dual-mode nanosensors usually use colorimetric and fluorescent reporters to achieve convenient visual inspection and highly sensitive fluorescent detection [160,161]. Wang et al. developed a dual-mode aptamer-based fluorescent sensor for the detection of PV, the major allergen of fish (Figure 3B) [159]. In Figure 3C(a), aptamer towards PV was obtained by in vitro screening of random ssDNA library containing a 40-mer randomized region using the triple-mode GO-SELEX. The aptamer-modified gold nanoparticle (AuNP-APT), complementary short-strand modified gold nanoparticles (AuNP-CS1), and fluorescent dye-labeled complementary short-strands (FAM-CS2) were assembled by DNA hybridization. After the addition of PV, the competitive interaction with aptamer leads to the decomposition of the aptamer sensor, resulting in the color shift of the AuNPs solution and the recovery of the FAM-CS2 fluorescence signal. The results showed that the aptamer sensor showed a good colorimetric response (2.5–20 $\mu g\ mL^{-1}$) and linear fluorescence correlation (2.38–40 $\mu g\ mL^{-1}$) in the PV concentration range. In addition, the affinity and specificity of the aptamer sensor were also investigated, as shown in Figure 3C(b,c). Therefore, aptamer 5 with good affinity (KD = 7.66 $\times$ 10^{-7} M) and specificity is the best aptamer for aptasensor construction. They also studied the feasibility of aptamer sensor in real fish samples, revealing the potential in field of monitoring and quantitative detection of food allergens.

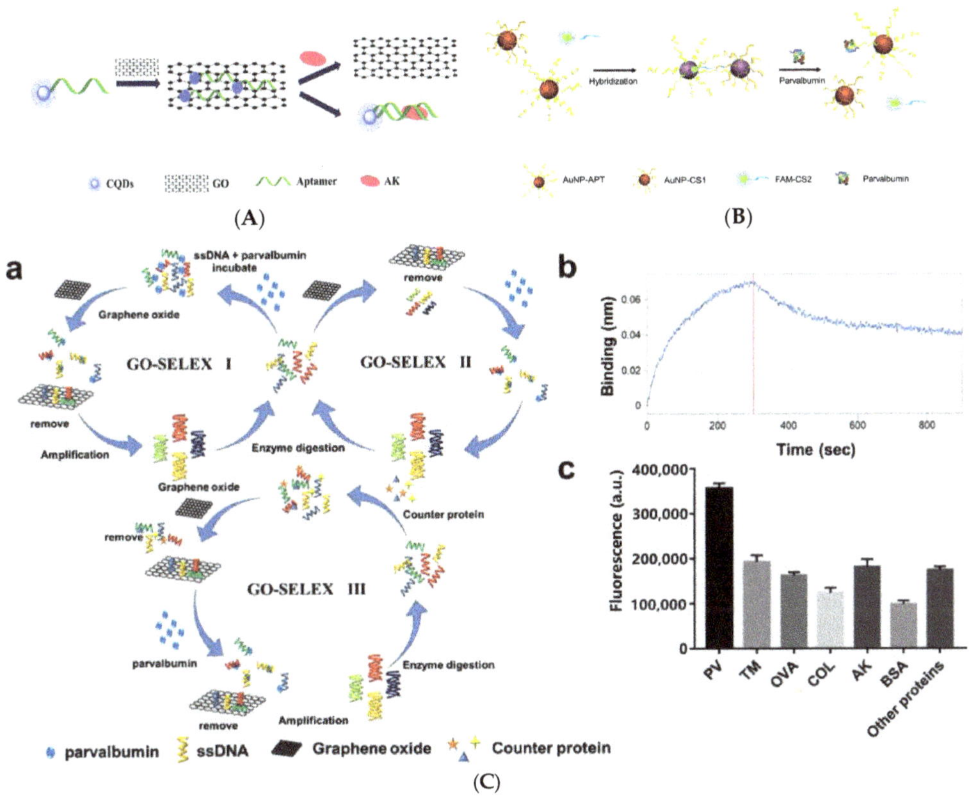

Figure 3. (**A**) Schematic of a "on-off-on" fluorescence aptasensor for AK detection. Reproduced with permission from [158]. Copyright Microchemical Journal, 2020. (**B**) Schematic of a dual-mode fluorescence sensor for PV detection based on AuNP color changes and FAM-CS2 fluorescence changes. (**C**) a: Schematic of the aptamer selection procedure by capturing GO-SELEX; b: Affinity of Apt5 towards PV; c: Specificity of Apt5 towards PV. Reproduced with permission from [159]. Copyright Microchemical Journal, 2020.

Recently, as an alternative to antibodies, the use of peptide aptamers as biosensors has attracted more attention. Peptide aptamers usually contain 10–20 amino acids, which the high selective recognition ability is equivalent to that of antibodies. Phadke et al. used ribosome display technology to select two fluorescent peptide aptamers Cas1 and Cas2. for the detection of α-casein [162]. Among them, 7-nitrobenzofurazan (NBD)-modified aminophenylalanine is coupled to the translated peptides to prepare fluorescent peptide aptamers. This is because the peptide can quench the fluorescence of NBD. Once the peptide recognizes the target α-casein, the NBD-modified phenylalanine is released, and its fluorescence will instantly increase. It is worth noting that although the fluorescence of the two aptamers increased slightly in the presence of the control protein β-lactoglobulin, the modification of Cas1 with polyethylene glycol (PEG-Cas1) inhibited this phenomenon, which is because PEG-Cas1 may inhibit the interaction between aptamer and β-lactoglobulin. The aptamer sensor with a LOD of 0.04 mM, is equivalent to that of the kit. Moreover, the system can detect α-casein in short time (20–25 s) when compared with the 15 min required by immunochromatography kits. In addition, it is found that when using PEG-Cas1 to detect casein, the instant increase in fluorescence can be observed even with the naked eye. This study has contributed to improving the specificity of aptamers.

In order to reduce the incidence of milk allergy, hypoallergenic formula (HF) has been commercialized as a substitute for milk [163]. Nevertheless, in some cases, infants who consume these formula milk powder still have allergic reactions because of residual β-lactoglobulin in HF [164]. Therefore, it is necessary to establish a method that can detect the lower concentration of β-lactoglobulin. Shi et al. used carbon dots (CDs) as a fluorescent signal and Fe_3O_4 NPs as a magnetic separator to establish a fluorescent-labeled assay for the detection of β-lactoglobulin [165]. The assay is based on the hybridization between aptamers immobilized on Fe_3O_4 NPs and CDs-labeled complementary oligonucleotides (cDNA). In the presence of β-lactoglobulin, the aptamer preferentially binds to β-lactoglobulin, and part of the CDs-cDNA is released into the solution. After magnetic separation, the fluorescence signal of the supernatant increased with the increase of β-lactoglobulin concentration. Based on this, the aptamer assay with the range of 0.25–50 ng mL^{-1} and a LOD of 37 pg mL^{-1} has been successfully applied to the detection of trace β-lactoglobulin in HF. In the study of Qi et al., the binding mechanism of aptamer and β-lactoglobulin and the detection principle of fluorescent surface-enhanced Raman scattering (fluorescent-SERS) dual-mode aptamer sensor were thoroughly studied, which provides a theory basis and application potential for the development of aptamer sensors [166]. In Figure 4a, the circular dichroism of Lg-18, thermodynamic parameters analysis, secondary structure of Lg-18, and the result of molecular docking between aptamer Lg-18 and β-lactoglobulin were performed to illustrate the successful selection of aptamer. The specific response principle of the dual-mode aptamer sensor is shown in Figure 4b. The fluorescent-SERS aptamer sensor shows a wider linear range (10–5000 ng mL^{-1}), and the LOD is 0.05 ng mL^{-1}. Furthermore, under the interference of other proteins, the aptamer sensor showed excellent specificity.

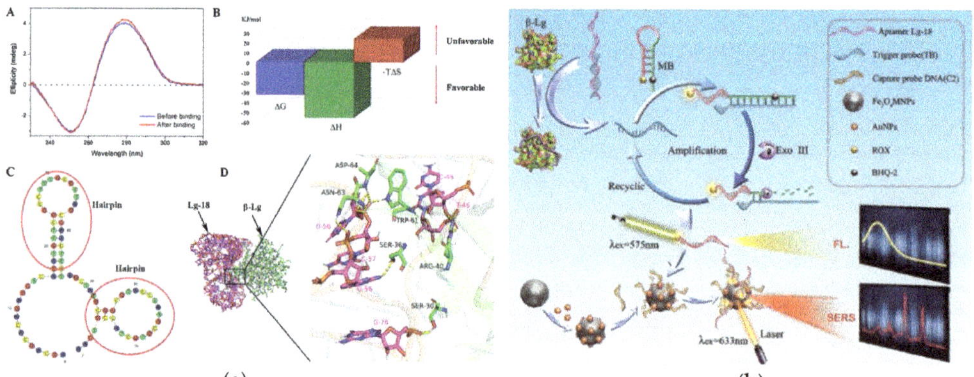

Figure 4. (**a**) Schematic of Lg-18 and β-lactoglobulin binding. A: Circular dichroism analysis of Lg-18 before and after binding. B: Analysis of thermodynamic parameters in the combined process. C: Secondary structure of Lg-18 predicted by Mfold online software. D: Analysis of molecular docking results of aptamer Lg-18 and β-lactoglobulin (**b**) Schematic of aptamer-based fluorescent Raman dual-mode biosensor for detection of β-lactoglobulin. Reproduced with permission from [166]. Copyright Sensors and Actuators B-Chemical, 2021.

Lys, as an allergen in egg and a biomarker of many diseases, its detection and quantification are of great significance in clinical diagnosis [167]. Sapkota et al. developed an aptamer sensor based on single-molecule FRET (smFRET) for the detection of Lys (Figure 5a) [168]. One of the arms has a blocking chain (B1), which is extended by 15 nucleotides to partially hybridize to the aptamer. The aptamer sensor remains open and almost no FRET efficiency occurs when Lys is not detected. After the addition of Lys, the aptamer binds to Lys and is displaced from the sensor, resulting in a foothold-mediated replacement of B1 by another chain, H1. At this time, the binding of Lys triggers the conformational transition state from low FRET to high FRET. Using this strategy, they demonstrated that the aptamer sensor can detect Lys at concentration as low as 30 nM, with a dynamic range extends to ~2 μM, and is almost free of interference from similar biomolecules. In addition, the smFRET method requires only a small number of aptamers, which offers the advantage of cost-effectiveness. In fluorescence detection, the presence of background fluorescence induced by ultraviolet-visible light in biological samples can lead to inaccurate detection results [169]. However, Ou et al. developed an X-ray nanocrystal scintillator aptamer sensor for sensitive detection of Lys based on the characteristics of weak scattering and almost no absorption of biological chromophores under X-ray irradiation (Figure 5b) [170]. In this study, aptamer-labeled lanthanide-doped nanocrystalline scintillators are designed to detect Lys quickly and sensitively through FRET. The use of low-dose X-rays as the excitation source and nanocrystals containing heavy atoms can achieve efficient luminescence, which endows the aptamer fluorescence sensor with high sensitivity (LOD: 0.94 nM), specificity, and sample recovery. In addition, this technology can provide a new generation of high-efficiency strategy without autofluorescence interference for the sensing and detection of biomarkers in biomedical applications.

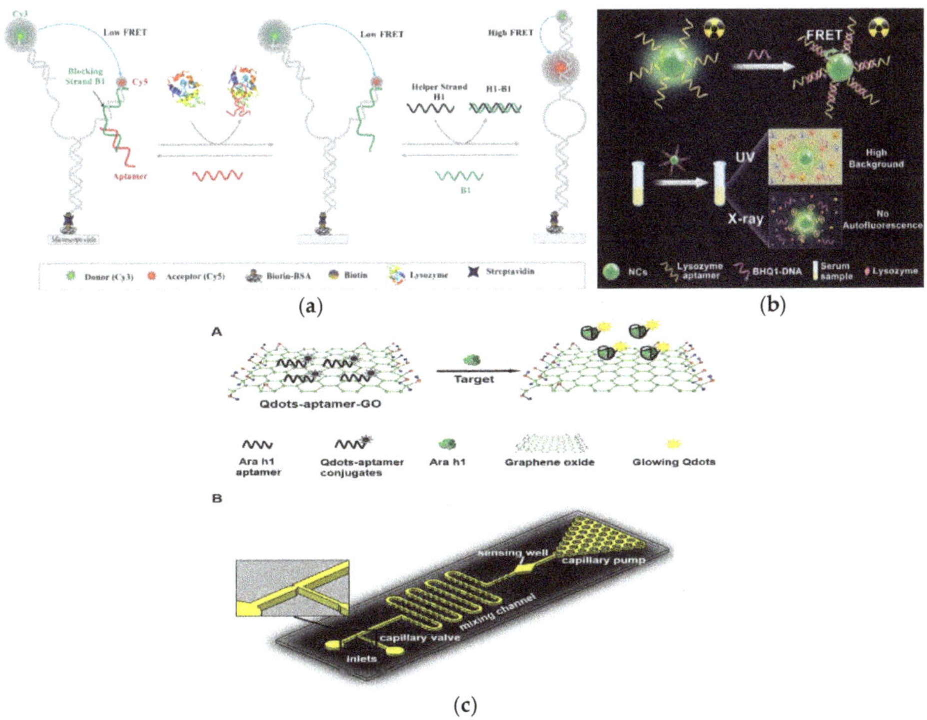

Figure 5. (**a**) Schematic of aptamer-based fluorescent biosensor for Lys detection. Reproduced with permission from [168]. Copyright sensors, 2020. (**b**) Schematic of aptamer sensor based on nanocrystal scintillator for detecting Lys without autofluorescence. Reproduced with permission from [170]. Copyright Analytical Chemistry, 2019. (**c**) A: Schematic of the Qdots-aptamer-GO quenching sensing principle; B: Schematic of the designed microfluidic chip. Reproduced with permission from [171]. Copyright Biosensors & Bioelectronics, 2016.

3.4. Detection of Plant Food Allergens

Lupin, a legume plant, is widely cultivated around the world. α-Conglutin and β-conglutin are the main allergens of lupin, accounting for 33% and 45%, respectively [172,173]. FRET signals based on ligand-induced conformational changes of the aptamer, Mairal et al. developed a dimer-based dual fluorescent carrier-labeled biosensor using 11 mer truncated aptamers for the sensitive detection of β-conglutin [174]. According to reports, the 11-mer truncated aptamer can form a dimer structure in its natural state [175]. Based on this, the FRET probe was prepared by ligating two fluorophores with 11 mer truncated aptamers. The fluorophore Alexa Fluor 488 is excited at 492 nm and emits at 519 nm, and Alexa Fluor 555 is excited at 533 nm and emits at 568 nm. After the addition of β-conglutin, specific interactions can cause changes in the structure of the double aptamer, resulting in increased fluorescence at 568 nm. This method not only has the advantages of strong specificity, rapid detection, and high sensitivity (LOD: 150 pM), but can also be used for the direct detection of β-conglutin in food matrices at room temperature.

FRET based fluorescence detection methods are receiving increasing attention due to their high sensitivity. Weng et al. also used this mechanism to develop a quantum dot (Qdots) aptamer functionalized GO nano-biosensor integrated microfluidic system for the detection of peanut allergen Ara h 1 (Figure 5c) [171]. The nano-biosensor uses the Qdots-aptamer-GO complex as a probe, which will undergo conformational changes when interacting with Ara h 1. This is because the correlation constant between Qdots-aptamer and Ara h1 is greater than that between Qdots-aptamer and GO, which causes Qdots-aptamer to be released from GO, thereby restoring fluorescence. This one-step "turn on" detection with a LOD of 56 ng mL^{-1} in the ready-to-use microfluidic chip only needs 10 min to achieve the sensitive detection of Ara h 1. In addition, the use of small optical detectors to measure fluorescent signals improves the portability of the entire system and provides a promising method for the rapid, accurate, and economical on-site detection of other food allergens.

4. Conclusions and Prospective

In the field of food safety, food allergy is a worldwide problem. Therefore, the development of effective allergen quantitative detection technology is a problem that the food industry has been exploring. Benefiting from the development of aptamer technology, different types of aptamer sensors can be established by using the highly specific binding of aptamers and targets. In addition, the detection technology based on fluorescent signal has the advantages of no radioactivity, simple operation, high throughput, high sensitivity, and small sample size, which have attracted more and more attention. Therefore, fluorescent biosensors using aptamers as biorecognition ligands have made great progress in the sensitive, rapid, specific, and simple analysis of food allergens. Furthermore, fluorescent biosensors can be integrated with inexpensive and portable equipment, which provides a theoretical basis and practical application experience for the portable detection of food allergens. For food manufacturers, portable devices with appropriate sensitivity and specificity are the best choice for allergen testing during food processing. It is worth noting that the fluorescent biosensor allows to capture different allergens in food samples by changing different aptamer, which provides a strategy for the wide application of aptamer-based fluorescent sensors.

In recent years, aptamer-based fluorescent sensors have been successfully applied in the detection of allergens and other small molecules. Although aptamers have strong specificity to the target, when other analogs coexist or the concentration of the target is low, the influence of this weak non-specific signal cannot be ignored. Therefore, improving the level of specific aptamer screening technology is one of the future study directions of aptamer technology specificity. With the continuous efforts of scientific researchers, we believe that we will be able to gradually overcome the technical difficulties of aptamer fluorescence sensors in practical applications and make them more widely used in environmental detection, health, and medical fields.

Author Contributions: L.H. coordinated and organized the writing of the entire manuscript and completed the Section 1, Section 2, Section 3.1 and part of Section 3.2; M.P. provided the idea and financial support of the research and completed the Section 4. X.X. completed the part of Section 3.2; K.L. checked the language and format of the manuscript; J.Y. and S.W. (Shan Wang) completed the part of Section 3.3; S.W. (Shuo Wang) finally checked the quality of the article. All authors have read and agreed to the published version of the manuscript.

Funding: This work was supported by the National Key Research and Development Program of China (No. 2017YFC1600402), National Natural Science Foundation of China (No. 31972147), Open Project Program of State Key Laboratory of Food Nutrition and Safety, Tianjin University of Science and Technology (No. SKLFNS-KF-202115) and Natural Science Foundation of the Inner Mongolia Autonomous Region (No. 2021MS03077).

Institutional Review Board Statement: Not applicable.

Informed Consent Statement: Not applicable.

Data Availability Statement: Not applicable.

Conflicts of Interest: The authors declare no conflict of interest.

References

1. Sathe, S.K.; Liu, C.Q.; Zaffran, V.D. Food allergy. *Annu. Rev. Food Sci. Technol.* **2016**, *7*, 191–220. [CrossRef]
2. Carter, C.A.; Frischmeyer-Guerrerio, P.A. The genetics of food allergy. *Curr. Allergy Asthm. Rep.* **2018**, *18*, 2. [CrossRef] [PubMed]
3. Anvari, S.; Miller, J.; Yeh, C.Y.; Davis, C.M. IgE-Mediated food allergy. *Clin. Rev. Allergy Immunol.* **2019**, *57*, 244–260. [CrossRef] [PubMed]
4. Zhou, J.R.; Qi, Q.Q.; Wang, C.; Qian, Y.F.; Liu, G.M.; Wang, Y.B.; Fu, L.L. Surface plasmon resonance (SPR) biosensors for food allergen detection in food matrices. *Biosens. Bioelectron.* **2019**, *142*, 111449. [CrossRef] [PubMed]
5. Gupta, M.; Cox, A.; Nowak-Wegrzyn, A.; Wang, J.L. Diagnosis of food allergy. *Immunol. Allergy Clin.* **2018**, *38*, 39–52. [CrossRef] [PubMed]
6. Quake, C.; Nadeau, K.C. The role of epigenetic mediation and the future of food allergy research. *Semin. Cell. Dev. Biol.* **2015**, *43*, 125–130. [CrossRef]
7. He, S.F.; Li, X.; Wu, Y.; Wu, S.D.; Wu, Z.H.; Yang, A.S.; Tong, P.; Yuan, J.L.; Gao, J.Y.; Chen, H.B. A novel sandwich enzyme-linked immunosorbent assay with covalently bound monoclonal antibody and gold probe for sensitive and rapid detection of bovine beta-lactoglobulin. *Anal. Bionanl. Chem.* **2018**, *410*, 3693–3703. [CrossRef] [PubMed]
8. Kim, M.J.; Park, S.B.; Kang, H.B.; Lee, K.M.; Kim, H.Y. Development of ultrafast PCR for rapid detection of buckwheat allergen DNA (fag e 1) in processed foods. *Food Control* **2021**, *130*, 108334. [CrossRef]
9. Xiong, W.L.; Parker, C.H.; Boo, C.C.; Fiedler, K.L. Comparison of allergen quantification strategies for egg, milk, and peanut in food using targeted LC-MS/MS. *Anal. Bioanal. Chem.* **2021**, *413*, 5755–5766. [CrossRef]
10. Yu, Z.W.; Wang, Y.Q.; Li, Z.X.; Pramod, S.N.; Zhang, L.J.; Lin, H. Development of ELISA method for detecting crustacean major allergen tropomyosin in processed food samples. *Food Anal. Method.* **2019**, *12*, 2719–2729. [CrossRef]
11. Liu, C.Q.; Zaffran, V.D.; Gupta, S.; Roux, K.H.; Sathe, S.K. Pecan (*Carya illinoinensis*) detection using a monoclonal antibody-based direct sandwich enzyme-linked immunosorbent assay. *LWT-Food Sci. Technol.* **2019**, *116*, 108516. [CrossRef]
12. Kim, M.J.; Kim, H.I.; Kim, J.H.; Suh, S.M.; Kim, H.Y. Rapid on-site detection of shrimp allergen tropomyosin using a novel ultrafast PCR system. *Food Sci. Biotechnol.* **2019**, *28*, 591–597. [CrossRef]
13. Linacero, R.; Sanchiz, A.; Ballesteros, I.; Cuadrado, C. Application of real-time PCR for tree nut allergen detection in processed foods. *Crit. Rev. Food Sci.* **2020**, *60*, 1077–1093. [CrossRef] [PubMed]
14. De Angelis, E.; Pilolli, R.; Monaci, L. Coupling SPE on-line pre-enrichment with HPLC and MS/MS for the sensitive detection of multiple allergens in wine. *Food Control* **2017**, *73*, 814–820. [CrossRef]
15. Gezer, R.G.; Liu, G.L.; Kokini, J.L. Development of a biodegradable sensor platform from gold coated zein nanophotonic films to detect peanut allergen, Ara h1, using surface enhanced raman spectroscopy. *Talanta* **2016**, *150*, 224–232. [CrossRef] [PubMed]
16. Mishra, G.K.; Barfidokht, A.; Tehrani, F.; Mishra, R.K. Food safety analysis using electrochemical biosensors. *Foods* **2018**, *7*, 141. [CrossRef]
17. Bragazzi, N.L.; Amicizia, D.; Panatto, D.; Tramalloni, D.; Valle, I.; Gasparini, R. Quartz-crystal microbalance (QCM) for public health: An overview of its applications. *Adv. Protein Chem. Str.* **2015**, *101*, 149–211.
18. Xia, X.H.; He, Q.; Dong, Y.; Deng, R.J.; Li, J.H. Aptamer-based homogeneous analysis for food control. *Curr. Anal. Chem.* **2020**, *16*, 4–13. [CrossRef]
19. Abbaspour, A.; Norouz-Sarvestani, F.; Noon, A.; Soltani, N. Aptamer-conjugated silver nanoparticles for electrochemical dual-aptamer-based sandwich detection of staphylococcus aureus. *Biosens. Bioelectron.* **2015**, *68*, 149–155. [CrossRef] [PubMed]
20. Yan, C.; Zhang, J.; Yao, L.; Xue, F.; Lu, J.F.; Li, B.G.; Chen, W. Aptamer-mediated colorimetric method for rapid and sensitive detection of chloramphenicol in food. *Food Chem.* **2018**, *260*, 208–212. [CrossRef]

21. Song, S.H.; Gao, Z.F.; Guo, X.; Chen, G.H. Aptamer-based detection methodology studies in food safety. *Food Anal. Methods* **2019**, *12*, 966–990. [CrossRef]
22. Liu, L.S.; Lu, X.L.; Zhao, Y.X. Aptamer-based strategies for cancer diagnosis and therapy. *J. Nanosci. Nanotechnol.* **2016**, *16*, 6611–6621. [CrossRef]
23. Gray, B.P.; Kelly, L.; Ahrens, D.P.; Barry, A.P.; Kratschmer, C.; Levy, M.; Sullenger, B.A. Tunable cytotoxic aptamer-drug conjugates for the treatment of prostate cancer. *Proc. Natl. Acad. Sci. USA* **2018**, *115*, 4761–4766. [CrossRef]
24. Cheng, L.; Zhang, J.; Lin, Y.; Wang, Q.; Zhang, X.X.; Ding, Y.H.; Cui, H.F.; Fan, H. An electrochemical molecular recognition-based aptasensor for multiple protein detection. *Anal. Biochem.* **2015**, *491*, 31–36. [CrossRef]
25. Markova, L.V.; Myshkin, N.K.; Makarenko, V.M. Fluorescence method for quick transformer oil monitoring. *Chem. Technol. Fuels Oils* **2016**, *52*, 194–202. [CrossRef]
26. Zienkiewicz, K.; Alche, J.D.; Zienkiewicz, A.; Tormo, A.; Castro, A.J. Identification of olive pollen allergens using a fluorescence-based 2D multiplex method. *Electrophoresis* **2015**, *36*, 1043–1050. [CrossRef]
27. Wang, T.; Zeng, L.H.; Li, D.L. A review on the methods for correcting the fluorescence inner-filter effect of fluorescence spectrum. *Appl. Spectrosc. Rev.* **2017**, *52*, 883–908. [CrossRef]
28. Fu, L.L.; Qian, Y.F.; Zhou, J.R.; Zheng, L.; Wang, Y.B. Fluorescence-based quantitative platform for ultrasensitive food allergen detection: From immunoassays to DNA sensors. *Compr. Rev. Food Sci. Saf.* **2020**, *19*, 3343–3364. [CrossRef]
29. Zhang, G.X.; Liu, Y.L.; Yang, M.; Huang, W.S.; Xu, J.H. An aptamer-based, fluorescent and radionuclide dual-modality probe. *Biochimie* **2020**, *171*, 55–62. [CrossRef]
30. Ma, P.F.; Guo, H.L.; Duan, N.; Ma, X.Y.; Yue, L.; Gu, Q.H.; Wang, Z.P. Label free structure-switching fluorescence polarization detection of chloramphenicol with truncated aptamer. *Talanta* **2021**, *230*, 122349. [CrossRef] [PubMed]
31. Alhadrami, H.A.; Chinnappan, R.; Eissa, S.; Rahamn, A.A.; Zourob, M. High affinity truncated DNA aptamers for the development of fluorescence based progesterone biosensors. *Anal. Biochem.* **2017**, *525*, 78–84. [CrossRef]
32. Liu, Y.X.; Zhao, Q. Direct fluorescence anisotropy assay for cocaine using tetramethylrhodamine-labeled aptamer. *Anal. Bioanal. Chem.* **2017**, *409*, 3993–4000. [CrossRef]
33. Dijkema, D.; Emons, J.A.M.; Van de Ven, A.A.J.M.; Elberink, J.O. Fish allergy: Fishing for novel diagnostic and therapeutic options. *Clin. Rev. Allergy Immunol.* **2020**. [CrossRef]
34. Wai, C.Y.Y.; Leung, N.Y.H.; Leung, P.S.C.; Chu, K.H. Modulating shrimp tropomyosin-mediated allergy: Hypoallergen DNA vaccines induce regulatory T cells to reduce hypersensitivity in mouse model. *Int. J. Mol. Sci.* **2019**, *20*, 4656. [CrossRef] [PubMed]
35. Xing, P.; Yu, H.Q.; Li, M.; Xiao, X.J.; Jiang, C.L.; Mo, L.H.; Zhang, M.; Yang, P.C.; Liu, Z.G. Characterization of arginine kinase, anovel allergen of dermatophagoides farinae (Der f 20). *Am. J. Transl. Res.* **2015**, *7*, 2815–2823. [PubMed]
36. Wasik, M.; Nazimek, K.; Nowak, B.; Askenase, P.W.; Bryniarski, K. Delayed-type hypersensitivity underlying casein allergy is suppressed by extracellular vesicles carrying miRNA-150. *Nutrients* **2019**, *11*, 907. [CrossRef] [PubMed]
37. Cong, Y.J.; Zhou, S.Y.; Li, L.F. Identification of the critical amino acid residues of immunoglobulin E and immunoglobulin G epitopes in alpha-lactalbumin by alanine scanning analysis. *J. Food Sci.* **2016**, *81*, T2597–T2603.
38. Surucu, O.; Abaci, S. Electrochemical determination of beta-lactoglobulin in whey proteins. *J. Food Meas. Charact.* **2020**, *14*, 11–19. [CrossRef]
39. Suprun, M.; Getts, R.; Grishina, G.; Suarez-Farinas, M.; Sampson, H.A. Epitope-specific antibody binding on major hen's egg-white allergen, ovomucoid, using novel multiplex immunoassay. *Allergy* **2019**, *74*, 256–257.
40. Bermingham, M.; Filep, S.; Reid-Black, K.; Thorpe, C.; Wuenschmann, S.; Chapman, M.; Hindley, J. Two-site immunoassays for the quantification of major egg allergens ovomucoid (Gal D 1) and ovalbumin (Gal D 2). *Allergy* **2019**, *74*, 438.
41. Hwang, H.S.; Kim, B.S.; Park, H.; Park, H.Y.; Choi, H.D.; Kim, H.H. Type and branched pattern of N-glycans and their structural effect on the chicken egg allergen ovotransferrin: A comparison with ovomucoid. *Glycoconj. J.* **2014**, *31*, 41–50. [CrossRef]
42. Tan, L.; Chong, K.W.; Goh, S.H. Lysozyme and mucolytics—The hidden allergen. *Singap. Med. J.* **2020**, *61*, 497. [CrossRef]
43. Ramesh, M.; Yuenyongviwat, A.; Konstantinou, G.N.; Lieberman, J.; Pascal, M.; Masilamani, M.; Sampson, H.A. Peanut T-cell epitope discovery: Ara h 1. *J. Allergy Clin. Immunol.* **2016**, *137*, 1764–1771. [CrossRef]
44. Shroba, J.; Barnes, C.; Nanda, M.; Dinakar, C.; Ciaccio, C. Ara h2 levels in dust from homes of individuals with peanut allergy and individuals with peanut tolerance. *Allergy Asthma Proc.* **2017**, *38*, 192–196. [CrossRef] [PubMed]
45. Pandey, A.K.; Varshney, R.K.; Sudini, H.K.; Pandey, M.K. An improved enzyme-linked immunosorbent assay (ELISA) based protocol using seeds for detection of five major peanut allergens Ara h 1, Ara h 2, Ara h 3, Ara h 6, and Ara h 8. *Front. Nutr.* **2019**, *6*, 68. [CrossRef] [PubMed]
46. Gomes, R.; Loureiro, C.; Pita, C.; Todo-Bom, A. Tri a 19 and wheat exercised-dependent urticaria. *Allergy* **2015**, *70*, 605–606.
47. Schmidt, M.A.; Hymowitz, T.; Herman, E.M. Breeding and characterization of soybean Triple Null; a stack of recessive alleles of Kunitz Trypsin Inhibitor, Soybean Agglutinin, and P34 allergen nulls. *Plant Breed.* **2015**, *134*, 310–315. [CrossRef]
48. Geng, T.; Stojsin, D.; Liu, K.; Schaalje, B.; Postin, C.; Ward, J.; Wang, Y.C.; Liu, Z.L.; Li, B.; Glenn, K. Natural variability of allergen levels in conventional soybeans: Assessing variation across north and south america from five production years. *J. Agric. Food Chem.* **2017**, *65*, 463–472. [CrossRef]
49. Archila, L.D.; Chow, I.T.; McGinty, J.W.; Renand, A.; Jeong, D.; Robinson, D.; Farrington, M.L.; Kwok, W.W. Ana o 1 and Ana o 2 cashew allergens share cross-reactive CD4(+) T cell epitopes with other tree nuts. *Clin. Exp. Allergy* **2016**, *46*, 871–883. [CrossRef]

50. Schein, C.; Teuber, S.S.; Cheng, H.; Grimm, C.C.; Maleki, S.J. Antibodies to the physiochemical-consensus sequence of Jug r 2 containing glutamine-rich repeats bind allergens in peanuts and other tree nuts. *J. Allergy Clin. Immunol.* **2013**, *131*, AB21. [CrossRef]
51. Iwan, M.; Vissers, Y.M.; Fiedorowicz, E.; Kostyra, H.; Kostyra, E.; Savelkoul, H.F.J.; Wichers, H.J. Impact of maillard reaction on immunoreactivity and allergenicity of the hazelnut allergen Cor a 11. *J. Agric. Food Chem.* **2011**, *59*, 7163–7171. [CrossRef]
52. Mattison, C.P.; Grimm, C.C.; Li, Y.C.; Chial, H.J.; McCaslin, D.R.; Chung, S.Y.; Bren-Mattison, Y.; Wasserman, R.L. Identification and characterization of Ana o 3 modifications on arginine-111 residue in heated cashew nuts. *J. Agric. Food Chem.* **2017**, *65*, 411–420. [CrossRef]
53. Sato, S.; Yamamoto, M.; Yanagida, N.; Ito, K.; Ohya, Y.; Imai, T.; Nagao, M.; Borres, M.P.; Moverare, R.; Ebisawa, M. Jug r 1 sensitization is important in walnut-allergic children and youth. *J. Aller. Clin. Immunol. Pract.* **2017**, *5*, 1784–1786. [CrossRef] [PubMed]
54. Blankestijn, M.A.; Jager, C.F.D.; Blom, W.M.; Otten, H.G.; de Jong, G.A.H.; Gaspari, M.; Houben, G.F.; Knulst, A.C.; Verhoeckx, K.C.M. A subset of walnut allergic adults is sensitized to walnut 11S globulin Jug r 4. *Clin. Exp. Allergy* **2018**, *48*, 1206–1213. [CrossRef]
55. Masthoff, L.J.N.; Mattsson, L.; Zuidmeer-Jongejan, L.; Lidholm, J.; Andersson, K.; Akkerdaas, J.H.; Versteeg, S.A.; Garino, C.; Meijer, Y.; Kentie, P.; et al. Sensitization to Cor a 9 and Cor a 14 is highly specific for a hazelnut allergy with objective symptoms in Dutch children and adults. *J. Allergy Clin. Immunol.* **2013**, *132*, 393–399. [CrossRef]
56. Kabasser, S.; Hafner, C.; Chinthrajah, S.; Sindher, S.B.; Kumar, D.; Kost, L.E.; Long, A.J.; Nadeau, K.C.; Breiteneder, H.; Bublin, M. Identification of Pru du 6 as a potential marker allergen for almond allergy. *Allergy* **2021**, *76*, 1463–1472. [CrossRef]
57. Fu, L.L.; Wang, C.; Zhu, Y.; Wang, Y.B. Seafood allergy: Occurrence, mechanisms and measures. *Trends Food Sci. Technol.* **2019**, *88*, 80–92. [CrossRef]
58. Carrera, M.; Pazos, M.; Gasset, M. Proteomics-based methodologies for the detection and quantification of seafood allergens. *Foods* **2020**, *9*, 1134. [CrossRef] [PubMed]
59. Prester, L. Seafood allergy, toxicity, and intolerance: A review. *J. Am. Coll. Nutr.* **2016**, *35*, 271–283. [CrossRef]
60. Lee, H.L.; Tang, M.M.; Bakhtiar, M.F.; Yadzir, Z.H.M.; Johar, A. Sensitization to local seafood allergens in adult patients with atopic dermatitis in malaysia. *Int. Arch. Allergy Immunol.* **2021**, *182*, 153–157. [CrossRef]
61. Li, J.L.; Wang, H.F.; Cheng, J.H. DNA, protein and aptamer-based methods for seafood allergens detection: Principles, comparisons and updated applications. *Crit. Rev. Food Sci.* **2021**. [CrossRef]
62. Sicherer, S.H.; Sampson, H.A. Food allergy: Epidemiology, pathogenesis, diagnosis, and treatment. *J. Allergy Clin. Immunol.* **2014**, *133*, 291–307. [CrossRef]
63. Khora, S.S. Seafood-associated shellfish allergy: A comprehensive review. *Immunol. Investig.* **2016**, *45*, 504–530. [CrossRef]
64. Davis, C.M.; Gupta, R.S.; Aktas, O.N.; Diaz, V.; Kamath, S.D.; Lopata, A.L. Clinical management of seafood allergy. *J. Aller. Clin. Immunol. Pract.* **2020**, *8*, 37–44. [CrossRef]
65. Ruethers, T.; Taki, A.C.; Johnston, E.B.; Nugraha, R.; Le, T.T.K.; Kalic, T.; McLean, T.R.; Kamath, S.D.; Lopata, A.L. Seafood allergy: A comprehensive review of fish and shellfish allergens. *Mol. Immunol.* **2018**, *100*, 28–57. [CrossRef] [PubMed]
66. Saptarshi, S.R.; Sharp, M.F.; Kamath, S.D.; Lopata, A.L. Antibody reactivity to the major fish allergen parvalbumin is determined by isoforms and impact of thermal processing. *Food Chem.* **2014**, *148*, 321–328. [CrossRef]
67. Zhang, X.P.; Li, Y.R.; Tao, Y.; Wang, Y.; Xu, C.H.; Lu, Y. A novel method based on infrared spectroscopic inception-resnet networks for the detection of the major fish allergen parvalbumin. *Food Chem.* **2021**, *337*, 127986. [CrossRef] [PubMed]
68. Kuehn, A.; Morisset, M.; Hilger, C. Fish allergens beyond beta-parvalbumin. *Rev. Francaise Allergol.* **2016**, *56*, 260–261. [CrossRef]
69. Gamboa, P.M.; Bartolome, B.; Lirio, E.G.; Cuesta-Herranz, J.; Pastor-Vargas, C. Aldolase a new crustacea allergen. *Ann. Allergy Asthma Immunol.* **2018**, *121*, 246–247. [CrossRef] [PubMed]
70. Lin, J.L.; Wan, Q.; Gao, A.J.; Jiang, C.L.; Liu, Y.L.; Liu, Z.G.; Yang, P.C.; Liu, X.Y. Characterization of a novel allergen Der f 25, homologous to triose-phosphate isomerase, from dermatophagoides farinae. *Int. J. Clin. Exp. Med.* **2016**, *9*, 10829–10837.
71. Wu, C.C.; Lee, C.H.; Tyan, Y.C.; Huang, E.S.; Yu, W.T.; Yu, H.S. Identification of pyruvate kinase 2 as a possible crab allergen and analysis of allergenic proteins in crabs consumed in Taiwan. *Food Chem.* **2019**, *289*, 413–418. [CrossRef]
72. Faber, M.A.; Pascal, M.; El Kharbouchi, O.; Sabato, V.; Hagendorens, M.M.; Decuyper, I.I.; Bridts, C.H.; Ebo, D.G. Shellfish allergens: Tropomyosin and beyond. *Allergy* **2017**, *72*, 842–848. [CrossRef] [PubMed]
73. Nugraha, R.; Kamath, S.D.; Johnston, E.; Karnaneedi, S.; Ruethers, T.; Lopata, A.L. Conservation analysis of B-cell allergen epitopes to predict clinical cross-reactivity between shellfish and inhalant invertebrate illergens. *Front. Immunol.* **2019**, *10*, 2676. [CrossRef] [PubMed]
74. Fei, D.X.; Liu, Q.M.; Chen, F.; Yang, Y.; Chen, Z.W.; Cao, M.J.; Liu, G.M. Assessment of the sensitizing capacity and allergenicity of enzymatic cross-linked arginine kinase, the crab allergen. *Mol. Nutr. Food Res.* **2016**, *60*, 1707–1718. [CrossRef]
75. Francis, F.; Doyen, V.; Debaugnies, F.; Mazzucchelli, G.; Caparros, R.; Alabi, T.; Blecker, C.; Haubruge, E.; Corazza, F. Limited cross reactivity among arginine kinase allergens from mealworm and cricket edible insects. *Food Chem.* **2019**, *276*, 714–718. [CrossRef] [PubMed]
76. Hong, J.Y.; Gao, Q.C.; Xiao, X.J.; Cao, H.; Yuan, R.Y.; Liu, Z.G.; Chen, T.Q. T cell epitope of arginine kinase with CpG co-encapsulated nanoparticles attenuates a shrimp allergen-induced Th2-bias food allergy. *Biosci. Biotechnol. Biochem.* **2019**, *84*, 804–814.

77. Han, T.J.; Liu, M.; Huan, F.; Li, M.S.; Xia, F.; Chen, Y.Y.; Chen, G.X.; Cao, M.J.; Liu, G.M. Identification and cross-reactivity analysis of sarcoplasmic-calcium-binding protein: A novel allergen in crassostrea angulata. *J. Agric. Food Chem.* **2020**, *68*, 5221–5231. [CrossRef]
78. Li, M.S.; Xia, F.; Liu, M.; He, X.R.; Chen, Y.Y.; Bai, T.L.; Chen, G.X.; Wang, L.; Cao, M.J.; Liu, G.M. Cloning, Expression, and epitope identification of myosin light chain 1: An allergen in mud crab. *J. Agric. Food Chem.* **2019**, *67*, 10458–10469. [CrossRef]
79. Yu, C.; Gao, X.; Lin, H.; Lin, H.; Zhang, Z.Y.; Khan, M.U.; Li, Y.H.; Chen, Y.; Li, Z.X. Identification and amino acid analysis of allergenic epitopes of a novel allergen paramyosin (Rap v 2) from rapana venosa. *J. Agric. Food Chem.* **2021**, *69*, 5381–5391. [CrossRef]
80. Villa, C.; Costa, J.; Oliveira, M.B.P.P.; Mafra, I. Bovine milk allergens: A comprehensive review. *Compr. Rev. Food Sci. Food Saf.* **2018**, *17*, 137–164. [CrossRef] [PubMed]
81. Pastor-Vargas, C.; Maroto, A.S.; Diaz-Perales, A.; Villaba, M.; Diaz, N.C.; Vivanco, F.; Cuesta-Herranz, J. Sensitive detection of major food allergens in breast milk: First gateway for allergenic contact during breastfeeding. *Allergy* **2015**, *70*, 1024–1027. [CrossRef] [PubMed]
82. Nehra, M.; Lettieri, M.; Dilbaghi, N.; Kumar, S.; Marrazza, G. Nano-biosensing platforms for detection of cow's milk allergens: An overview. *Sensors* **2020**, *20*, 32. [CrossRef] [PubMed]
83. Chakraborty, G.; Ray, A.K.; Singh, P.K.; Pal, H. An exceptionally intense turn-on fluorescence sensor in the far-red region for common milk allergen, beta-lactoglobulin. *Sensor Actuators B Chem.* **2021**, *327*, 128864. [CrossRef]
84. Ashley, J.; Shukor, Y.; D'Aurelio, R.; Trinh, L.; Rodgers, T.L.; Temblay, J.; Pleasants, M.; Tothill, I.E. Synthesis of molecularly imprinted polymer nanoparticles for alpha-casein detection using surface plasmon resonance as a milk allergen sensor. *ACS Sensors* **2018**, *3*, 418–424. [CrossRef]
85. Ramachandran, B.; Yang, C.T.; Downs, M.L. Parallel reaction monitoring mass spectrometry method for detection of both casein and whey milk allergens from a baked food matrix. *J. Proteome Res.* **2020**, *19*, 2964–2976. [CrossRef]
86. Xu, N.F.; Wang, Y.J.; Pan, L.; Wei, X.L.; Wang, Y.F. Dual-labelled immunoassay with goldmag nanoparticles and quantum dots for quantification of casein in milk. *Food Agric. Immunol.* **2017**, *28*, 1105–1115. [CrossRef]
87. Ricciardi, C.; Santoro, K.; Stassi, S.; Lamberti, C.; Giuffrida, M.G.; Arlorio, M.; Decastelli, L. Microcantilever resonator arrays for immunodetection of beta-lactoglobulin milk allergen. *Sensor. Actuators B Chem.* **2018**, *254*, 613–617. [CrossRef]
88. Yang, A.S.; Zheng, Y.C.; Long, C.Y.; Chen, H.B.; Liu, B.; Li, X.; Yuan, J.L.; Cheng, F.F. Fluorescent immunosorbent assay for the detection of alpha-lactalbumin in dairy products with monoclonal antibody bioconjugated with CdSe/ZnS quantum dots. *Food Chem.* **2014**, *150*, 73–79. [CrossRef]
89. Montiel, V.R.V.; Campuzano, S.; Torrente-Rodriguez, R.M.; Reviejo, A.J.; Pingarron, J.M. Electrochemical magnetic beads-based immunosensing platform for the determination of alpha-lactalbumin in milk. *Food Chem.* **2016**, *213*, 595–601. [CrossRef] [PubMed]
90. Varlamova, E.G.; Zaripov, O.G. Beta-lactoglobulin-nutrition allergen and nanotransporter of different nature ligands therapy with therapeutic action. *Res. Vet. Sci.* **2020**, *133*, 17–25. [CrossRef] [PubMed]
91. Tan, J.W.; Joshi, P. Egg allergy: An update. *J. Paediatr. Child Health* **2014**, *50*, 11–15. [CrossRef]
92. Dona, D.W.; Suphioglu, C. Egg allergy: Diagnosis and immunotherapy. *Int. J. Mol. Sci.* **2020**, *21*, 5010. [CrossRef]
93. Sogawa, K.; Takahashi, Y.; Shibata, Y.; Satoh, M.; Kodera, Y.; Nomura, F.; Tanaka, T.; Sato, H.; Yamaide, F.; Nakano, T.; et al. Search for a novel allergen in hen's egg allergy using an IgE immunoblotting assay. *Int. Arch. Allergy Immunol.* **2018**, *176*, 189–197. [CrossRef] [PubMed]
94. Yanagida, N.; Sato, S.; Takahashi, K.; Asaumi, T.; Nagakura, K.; Ogura, K.; Takamatsu, N.; Ebisawa, M. Safe egg yolk consumption after a negative result for low-dose egg oral food challenge. *Pediatric Allerg. Immunol.-UK* **2021**, *32*, 170–176. [CrossRef] [PubMed]
95. Dhanapala, P.; Withanage-Dona, D.; Tang, M.L.K.; Doran, T.; Suphioglu, C. Hypoallergenic variant of the major egg white allergen gal d 1 produced by disruption of cysteine bridges. *Nutrients* **2017**, *9*, 171. [CrossRef] [PubMed]
96. Dang, T.D.; Peters, R.L.; Koplin, J.J.; Dharmage, S.C.; Gurrin, L.C.; Ponsonby, A.L.; Martino, D.J.; Neeland, M.; Tang, M.L.K.; Allen, K.J. Egg allergen specific IgE diversity predicts resolution of egg allergy in the population cohort HealthNuts. *Allergy* **2019**, *74*, 318–326. [CrossRef]
97. Pablos-Tanarro, A.; Lozano-Ojalvo, D.; Molina, E.; Lopez-Fandino, R. Assessment of the allergenic potential of the main egg white proteins in BALB/c mice. *J. Agric. Food Chem.* **2018**, *66*, 2970–2976. [CrossRef]
98. Benede, S.; Lopez-Fandino, R.; Reche, M.; Molina, E.; Lopez-Exposito, I. Influence of the carbohydrate moieties on the immunoreactivity and digestibility of the egg allergen ovomucoid. *PLoS ONE* **2013**, *8*, e80810. [CrossRef]
99. Kido, J.; Matsumoto, T. Attenuated allergenic activity of ovomucoid after electrolysis. *Allergy Asthma Immunol.* **2015**, *7*, 599–604. [CrossRef]
100. Lin, Y.T.; Wu, C.T.; Huang, J.L.; Cheng, J.H.; Yeh, K.W. Correlation of ovalbumin of egg white components with allergic diseases in children. *J. Microbiol. Immunol.* **2016**, *49*, 112–118. [CrossRef]
101. Dantas, M.D.D.; Tenorio, H.D.; Lopes, T.I.B.; Pereira, H.J.V.; Marsaioli, A.J.; Figueiredo, I.M.; Santos, J.C.C. Interactions of tetracyclines with ovalbumin, the main allergen protein from egg white: Spectroscopic and electrophoretic studies. *Int. J. Biol. Macromol.* **2017**, *102*, 505–514. [CrossRef] [PubMed]
102. Mukae, T.; Yoshii, K.; Watanobe, T.; Tagami, T.; Oishi, I. Production and characterization of eggs from hens with ovomucoid gene mutation. *Poult. Sci.* **2021**, *100*, 452–460. [CrossRef]

103. Tong, P.; Gao, L.; Gao, J.Y.; Li, X.; Wu, Z.H.; Yang, A.S.; Chen, H.B. Iron-induced chelation alleviates the potential allergenicity of ovotransferrin in a BALB/c mouse model. *Nutr. Res.* **2017**, *47*, 81–89. [CrossRef]
104. Tong, P.; Gao, J.Y.; Chen, H.B.; Li, X.; Zhang, Y.; Jian, S. Preparation and immunological reactions of a purified egg allergen ovotransferrin. *Int. J. Food Prop.* **2014**, *17*, 293–308. [CrossRef]
105. Chen, L.P.; Xia, N.; Li, T.T.; Bai, Y.H.; Chen, X.L. Aptasensor for visual and fluorometric determination of lysozyme based on the inner filter effect of gold nanoparticles on CdTe quantum dots. *Microchim. Acta* **2016**, *183*, 2917–2923. [CrossRef]
106. Boushell, V.; Pang, S.; He, L.L. Aptamer-based SERS detection of lysozyme on a food-handling surface. *J. Food Sci.* **2017**, *82*, 225–231. [CrossRef]
107. Pi, X.W.; Wan, Y.; Yang, Y.L.; Li, R.Y.; Wu, X.J.; Xie, M.Y.; Li, X.; Fu, G.M. Research progress in peanut allergens and their allergenicity reduction. *Trends Food Sci. Technol.* **2019**, *93*, 212–220. [CrossRef]
108. Palladino, C.; Breiteneder, H. Peanut allergens. *Mol. Immunol.* **2018**, *100*, 58–70. [CrossRef]
109. Bublin, M.; Breiteneder, H. Cross-reactivity of peanut allergens. *Curr. Allergy Asthm. Rep.* **2014**, *14*, 426. [CrossRef] [PubMed]
110. Filep, S.; Block, D.S.; Smith, B.R.E.; King, E.M.; Commins, S.; Kulis, M.; Vickery, B.P.; Chapman, M.D. Specific allergen profiles of peanut foods and diagnostic or therapeutic allergenic products. *J. Allergy Clin. Immunol.* **2018**, *141*, 626–631. [CrossRef] [PubMed]
111. Wu, Z.H.; Zhou, N.L.; Xiong, F.Q.; Li, X.; Yang, A.S.; Tong, P.; Tang, R.H.; Chen, H.B. Allergen composition analysis and allergenicity assessment of Chinese peanut cultivars. *Food Chem.* **2016**, *196*, 459–465. [CrossRef]
112. Guillon, B.; Bernard, H.; Drumare, M.F. Heat processing of peanut seed enhances the sensitization potential of the major peanut allergen Ara h 6. *Mol. Nutr. Food Res.* **2016**, *60*, 2722–2735. [CrossRef]
113. Yin, H.Y.; Fang, T.J.; Li, Y.T.; Fung, Y.F.; Tsai, W.C.; Dai, H.Y.; Wen, H.W. Rapidly detecting major peanut allergen-Ara h2 in edible oils using a new immunomagnetic nanoparticle-based lateral flow assay. *Food Chem.* **2019**, *271*, 505–515. [CrossRef]
114. Ramesh, M.; Masilamani, M.; Konstantinou, G.N.; Lieberman, J.A.; Sampson, H.A.; Sivakumar, M.; Yuenyongviwat, A.; Pascal, M. Peanut T cell epitope discovery: Ara h1 and Ara h3. *J. Allergy Clin. Immunol.* **2014**, *133*, AB233. [CrossRef]
115. Zhao, L.P.; Zhao, L.; Zhang, B.C.; Robotham, J.M.; Roux, K.H.; Tang, H.L. Identification of a common Ara h 3 epitope recognized by both the capture and the detection monoclonal antibodies in an ELISA detection kit. *PLoS ONE* **2017**, *12*, e0182935. [CrossRef] [PubMed]
116. Keser, M.; Gummadov, N.; Akin, B.; Belen, S.; Mert, Z.; Taner, S.; Topal, A.; Yazar, S.; Morgounov, A.; Sharma, R.C.; et al. Genetic gains in wheat in Turkey: Winter wheat for dryland conditions. *Crop J.* **2017**, *5*, 533–540. [CrossRef]
117. Pequeno, D.N.L.; Hernandez-Ochoa, I.M.; Reynolds, M.; Sonder, K.; MoleroMilan, A.; Robertson, R.D.; Lopes, M.S.; Xiong, W.; Kropff, M.; Asseng, S. Climate impact and adaptation to heat and drought stress of regional and global wheat production. *Environ. Res. Lett.* **2021**, *16*, 054070. [CrossRef]
118. Brouns, F.R.; van Rooy, G.; Shewry, P.; Rustgi, S.; Jonkers, D. Adverse reactions to wheat or wheat components. *Compr. Rev. Food Sci. Food Saf.* **2019**, *18*, 1437–1452. [CrossRef] [PubMed]
119. Wassmann, A.; Schafer, C. Wheat allergy: A challenging diagnosis for physicians and nutritionists. *Allergologie* **2014**, *37*, 466–475.
120. Baatjies, R.; Meijster, T.; Heederik, D.; Jeebhay, M.F. Exposure-response relationships for inhalant wheat allergen exposure and asthma. *Occup. Environ. Med.* **2015**, *72*, 200–207. [CrossRef]
121. Muntean, I.A.; Bocsan, I.C.; Deleanu, D. Oral Immunotherapy to wheat in allergic asthmatic female: A case report. *Iran. J. Allergy Asthm.* **2018**, *17*, 291–294.
122. Takemoto, K.; Asano, M.; Takahashi, K. Identification and and characterization of the major allergens in whole wheat norin 61. *J. Jpn. Soc. Food. Sci.* **2015**, *62*, 104–113. [CrossRef]
123. Pacharn, P.; Kumjim, S.; Tattiyapong, P.; Jirapongsananuruk, O.; Piboonpocanun, S. Identification of wheat sensitization using an in-house wheat extract On Coca-10% alcohol solution children with wheat anaphylaxis. *Asian Pac. J. Allergy* **2016**, *34*, 153–158.
124. Hofer, G.; Wieser, S.; Bogdos, M.K.; Gattinger, P.; Nakamura, R.; Ebisawa, M.; Makela, M.; Papadopoulos, N.; Valenta, R.; Keller, W. Three-dimensional structure of the wheat-amylase Tri a 17, a clinically relevant food allergen. *Allergy* **2019**, *74*, 1009–1013. [CrossRef]
125. Basaj, A.; Kennard, L.; Mirakian, R.; Wagner, A. Patients sensitised to the wheat lipid transfer protein Tri a 14 do not cross-react with omega-5 gliadin. *Clin. Exp. Allergy* **2019**, *49*, 1689.
126. Baar, A.; Pahr, S.; Constantin, C.; Giavi, S.; Manoussaki, A.; Papadopoulos, N.G.; Ebner, C.; Mari, A.; Vrtala, S.; Valenta, R. Specific IgE reactivity to Tri a 36 in children with wheat food allergy. *J. Allergy Clin. Immunol.* **2014**, *133*, 585–587. [CrossRef]
127. Matsuo, A.; Matsushita, K.; Fukuzumi, A.; Tokumasu, N.; Yano, E.; Zaima, N.; Moriyama, T. Comparison of various soybean allergen levels in genetically and non-genetically modified soybeans. *Foods* **2020**, *9*, 522. [CrossRef]
128. Yoshimitsu, M.; Kiyota, K.; Kajimura, K.; Yamano, T. Development of an LC-MS/MS-based analytical method for quantification of soybean allergen Gly m 4 in soybean grains and processed foods. *Food Agric. Immunol.* **2019**, *30*, 25–33. [CrossRef]
129. Murakami, H.; Ogawa, T.; Takafuta, A.; Yano, E.; Zaima, N.; Moriyama, T. Identification of the 7S and 11S globulins as percutaneously sensitizing soybean allergens as demonstrated through epidermal application of crude soybean extract. *Biosci. Biotechnol. Biochem.* **2018**, *82*, 1408–1416. [CrossRef] [PubMed]
130. Bhunia, R.K.; Chakraborty, A.; Kaur, R.; Gayatri, T.; Bhattacharyya, J.; Basu, A.; Maiti, M.K.; Sen, S.K. Seed-specific increased expression of 2S albumin promoter of sesame qualifies it as a useful genetic tool for fatty acid metabolic engineering and related transgenic intervention in sesame and other oil seed crops. *Plant Mol. Biol.* **2014**, *86*, 351–365. [CrossRef] [PubMed]

131. Ueberham, E.; Spiegel, H.; Havenith, H.; Rautenberger, P.; Lidzba, N.; Schillberg, S.; Lehmann, J. Simplified tracking of a soy allergen in processed food using a monoclonal antibody-based sandwich ELISA targeting the soybean 2S albumin Gly m 8. *J. Agric. Food Chem.* **2019**, *67*, 8660–8667. [CrossRef]
132. Yang, H.; Qu, Y.Z.; Li, J.T.; Liu, X.Q.; Wu, R.N.; Wu, J.R. Improvement of the protein quality and degradation of allergens in soybean meal by combination fermentation and enzymatic hydrolysis. *LWT-Food Sci. Technol.* **2020**, *128*, 109442. [CrossRef]
133. Wang, W.; Zhu, X.D.; Teng, S.; Fan, Q.J.; Qian, H. Label-free biochips for rapid detection of soybean allergen GlymBd 30K (P34) in foods. *Trop. J. Pharm. Res.* **2017**, *16*, 755–760. [CrossRef]
134. Xi, J.; Shi, Q.Q. Development of an indirect competitive ELISA kit for the detection of soybean allergenic protein Gly m Bd 28K. *Food Anal. Methods* **2016**, *9*, 2998–3005. [CrossRef]
135. Xi, J.; He, M.X. Location of destroyed antigenic sites of Gly m Bd 60 K after three processing technologies. *Food Res. Int.* **2020**, *134*, 109199. [CrossRef]
136. Yilmaz, C.; Tas, N.G.; Kocadagli, T.; Gokmen, V. Determination of serotonin in nuts and nut containing products by liquid chromatography tandem mass spectrometry. *Food Chem.* **2019**, *272*, 347–353. [CrossRef]
137. Hummel, M.; Wigger, T.; Hoper, T.; Westkamp, I.; Brockmeyer, J. Simple, rapid, and selective isolation of 2S albumins from allergenic seeds and nuts. *J. Agric. Food Chem.* **2015**, *63*, 6035–6040. [CrossRef]
138. Dreskin, S.C.; Koppelman, S.J.; Andorf, S.; Nadeau, K.C.; Kalra, A.; Braun, W.; Negi, S.S.; Chen, X.N.; Schein, C.H. The importance of the 2S albumins for allergenicity and cross-reactivity of peanuts, tree nuts, and sesame seeds. *J. Allergy Clin. Immunol.* **2021**, *147*, 1154–1163. [CrossRef] [PubMed]
139. Reitsma, M.; Bastiaan-Net, S.; Sforza, S.; van der Valk, J.P.M.; van Wijk, R.V.; Savelkoul, H.F.J.; de Jong, N.W.; Wichers, H.J. Purification and characterization of anacardium occidentale (cashew) allergens Ana o 1, Ana o 2, and Ana o 3. *J. Agric. Food Chem.* **2016**, *64*, 1191–1201. [CrossRef]
140. Chen, L.; Rashid, F.; Shah, A.; Awan, H.M.; Wu, M.M.; Liu, A.; Wang, J.; Zhu, T.; Luo, Z.F.; Shan, G. The isolation of an RNA aptamer targeting to p53 protein with single amino acid mutation. *Proc. Natl. Acad. Sci. USA* **2015**, *112*, 10002–10007. [CrossRef]
141. Zhao, L.P.; Yang, G.; Zhang, X.M.; Qu, F. Development of apatmer screening against proteins and its applications. *Chin. J. Anal. Chem.* **2020**, *48*, 560–572. [CrossRef]
142. Kohlberger, M.; Gadermaier, G. SELEX: Critical factors and optimization strategies for successful aptamer selection. *Biotechnol. Appl. Biochem.* **2021**. [CrossRef]
143. Daems, E.; Moro, G.; Campos, R.; De Wael, K. Mapping the gaps in chemical analysis for the characterisation of aptamer-target interactions. *TrAC-Trends Anal. Chem.* **2021**, *142*, 116311. [CrossRef]
144. Entzian, C.; Schubert, T. Studying small molecule-aptamer interactions using MicroScale Thermophoresis (MST). *Methods* **2016**, *97*, 27–34. [CrossRef]
145. Drabik, A.; Ner-Kluza, J.; Mielczarek, P.; Civit, L.; Mayer, G.; Silberring, J. Advances in the study of aptamer-protein target identification using the chromatographic approach. *J. Proteome Res.* **2018**, *17*, 2174–2181. [CrossRef] [PubMed]
146. Arib, C.; Liu, Q.Q.; Djaker, N.; Fu, W.L.; de la Chapelle, M.L.; Spadavecchia, J. Influence of the aptamer grafting on its conformation and its interaction with targeted protein. *Plasmonics* **2019**, *14*, 1029–1038. [CrossRef]
147. Zhou, C.; Xin, L.; Duan, X.Y.; Urban, M.J.; Liu, N. Dynamic plasmonic system that responds to thermal and aptamer-target regulations. *Nano Lett.* **2018**, *18*, 7395–7399. [CrossRef]
148. Szameit, K.; Berg, K.; Kruspe, S.; Valentini, E.; Magbanua, E.; Kwiatkowski, M.; de Beauchene, I.C.; Krichel, B.; Schamoni, K.; Uetrecht, C.; et al. Structure and target interaction of a G-quadruplex RNA-aptamer. *RNA Biol.* **2016**, *13*, 973–987. [CrossRef]
149. Zhao, Q.; Tao, J.F.; Feng, W.; Uppal, J.S.; Peng, H.Y.; Le, X.C. Aptamer binding assays and molecular interaction studies using fluorescence anisotropy—A review. *Anal. Chim. Acta* **2020**, *1125*, 267–278. [CrossRef]
150. Soh, J.H.; Lin, Y.Y.; Rana, S.; Ying, J.Y.; Stevens, M.M. Colorimetric detection of small molecules in complex matrixes via target-mediated growth of aptamer-functionalized gold nanoparticles. *Anal. Chem.* **2015**, *87*, 7644–7652. [CrossRef] [PubMed]
151. Bi, W.J.; Bai, X.; Gao, F.; Lu, C.C.; Wang, Y.; Zhai, G.J.; Tian, S.S.; Fan, E.G.; Zhang, Y.K.; Zhang, K. DNA-templated aptamer probe for identification of target proteins. *Anal. Chem.* **2017**, *89*, 4071–4076. [CrossRef]
152. Ilgu, M.; Yan, S.T.; Khounlo, R.M.; Lamm, M.H.; Nilsen-Hamilton, M. Common secondary and tertiary structural features of aptamer-ligand interaction shared by RNA aptamers with different primary sequences. *Molecules* **2019**, *24*, 4535. [CrossRef]
153. Tong, W.S.; Yuen, A.W.T.; Wai, C.Y.Y.; Leung, N.Y.H.; Chu, K.H.; Leung, P.S.C. Diagnosis of fish and shellfish allergies. *J. Asthma Allergy.* **2018**, *11*, 247–260. [CrossRef]
154. Zhang, Y.X.; Wu, Q.P.; Sun, M.; Zhang, J.M.; Mo, S.P.; Wang, J.; Wei, X.H.; Bai, J.L. Magnetic-assisted aptamer-based fluorescent assay for allergen detection in food matrix. *Sensor. Actuators B-Chem.* **2018**, *263*, 43–49. [CrossRef]
155. Zhang, Y.X.; Wu, Q.P.; Wei, X.H.; Zhang, J.M.; Mo, S.P. DNA aptamer for use in a fluorescent assay for the shrimp allergen tropomyosin. *Microchim. Acta* **2017**, *184*, 633–639. [CrossRef]
156. Chinnappan, R.; Rahamn, A.A.; AlZabn, R.; Kamath, S.; Lopata, A.L.; Abu-Salah, K.M.; Khalid, M.; Zourob, M. Aptameric biosensor for the sensitive detection of major shrimp allergen, Tropomyosin. *Food Chem.* **2020**, *314*, 126133. [CrossRef]
157. Shokoufi, N.; Heshi, S.V. Enhancmentation of photo-thermal lens of fluorescence molecules by fluorescence resonance energy transfer mechanism. *J. Fluoresc.* **2021**, *31*, 587–593. [CrossRef] [PubMed]

158. Zhou, J.R.; Ai, R.; Weng, J.J.; Li, L.F.; Zhou, C.S.; Ma, A.J.; Fu, L.L.; Wang, Y.B. A "on-off-on" fluorescence aptasensor using carbon quantum dots and graphene oxide for ultrasensitive detection of the major shellfish allergen arginine kinase. *Microchem. J.* **2020**, *158*, 105171. [CrossRef]
159. Wang, Y.B.; Li, H.; Zhou, J.R.; Qi, Q.Q.; Fu, L.L. A colorimetric and fluorescent gold nanoparticle-based dual-mode aptasensor for parvalbumin detection. *Microchem. J.* **2020**, *159*, 105413. [CrossRef]
160. Shi, Y.P.; Pan, Y.; Zhang, H.; Zhang, Z.M.; Li, M.J.; Yi, C.Q.; Yang, M.S. A dual-mode nanosensor based on carbon quantum dots and gold nanoparticles for discriminative detection of glutathione in human plasma. *Biosens. Bioelectron.* **2014**, *56*, 39–45. [CrossRef]
161. Liu, T.; Li, N.; Dong, J.X.; Zhang, Y.; Fan, Y.Z.; Lin, S.M.; Luo, H.Q.; Li, N.B. A colorimetric and fluorometric dual-signal sensor for arginine detection by inhibiting the growth of gold nanoparticles/carbon quantum dots composite. *Biosens. Bioelectron.* **2017**, *87*, 772–778. [CrossRef] [PubMed]
162. Phadke, C.; Tada, S.; Kono, I.; Hiyama, A.; Takase, Y.; Gayama, S.; Aigaki, T.; Ito, Y.; Uzawa, T. Instantaneous detection of alpha(s)-casein in cow's milk using fluorogenic peptide aptamers. *Anal. Methods-UK* **2020**, *12*, 1368–1373. [CrossRef]
163. Lozano-Ojalvo, D.; Perez-Rodriguez, L.; Pablos-Tanarro, A.; Lopez-Fandino, R.; Molina, E. Pepsin treatment of whey proteins under high pressure produces hypoallergenic hydrolysates. *Innov. Food Sci. Emerg.* **2017**, *43*, 154–162. [CrossRef]
164. Shi, M.L.; Cen, Y.; Sohail, M.; Xu, G.H.; Wei, F.D.; Ma, Y.S.; Xu, X.M.; Ma, Y.J.; Song, Y.Y.; Hu, Q. Aptamer based fluorometric beta-lactoglobulin assay based on the use of magnetic nanoparticles and carbon dots. *Microchim. Acta* **2018**, *185*, 40. [CrossRef] [PubMed]
165. Qi, S.; Duan, N.; Sun, Y.H.; Zhou, Y.; Ma, P.F.; Wu, S.J.; Wang, Z.P. High-affinity aptamer of allergen beta-lactoglobulin: Selection, recognition mechanism and application. *Sensor Actuators B-Chem.* **2021**, *340*, 129956. [CrossRef]
166. Panahipour, L.; Tabatabaei, A.A.; Gruber, R. Hypoallergenic infant formula lacks transforming growth factor beta activity and has a lower anti-inflammatory activity than regular infant formula. *J. Dairy Sci.* **2020**, *103*, 6771–6781. [CrossRef]
167. Wu, J.; Hou, Y.; Wang, P.Y.; Wang, Z.N.; Li, Y.J.; Wang, S.; Yang, M. Detection of lysozyme with aptasensor based on fluorescence resonance energy transfer from carbon dots to graphene oxide. *Luminescence* **2016**, *31*, 1207–1212. [CrossRef]
168. Sapkota, K.; Dhakal, S. FRET-based aptasensor for the selective and sensitive detection of lysozyme. *Sensors* **2020**, *20*, 914. [CrossRef]
169. Li, X.L.; Yu, Z.S. Determination of selenium in biological samples with an energy -dispersive X-ray fluorescence spectrometer. *Appl. Radiat. Isot.* **2016**, *111*, 45–49. [CrossRef]
170. Ou, X.Y.; Chen, Y.Y.; Xie, L.L.; Chen, J.; Zan, J.; Chen, X.F.; Hong, Z.Z.; He, Y.; Li, J.; Yang, H.H. X-ray nanocrystal scintillator-based aptasensor for autofluorescence-free detection. *Anal. Chem.* **2019**, *19*, 10149–10155. [CrossRef]
171. Jimenez-Lopez, J.C.; Foley, R.C.; Brear, E.; Clarke, V.C.; Lima-Cabello, E.; Florido, J.F.; Singh, K.B.; Alche, J.D.; Smith, P.M.C. Characterization of narrow-leaf lupin (*Lupinus angustifolius* L.) recombinant major allergen IgE-binding proteins and the natural beta-conglutin counterparts in sweet lupin seed species. *Food Chem.* **2018**, *244*, 60–70. [CrossRef] [PubMed]
172. Villa, C.; Costa, J.; Mafra, I. Lupine allergens: Clinical relevance, molecular characterization, cross-reactivity, and detection strategies. *Compr. Rev. Food Sci. Food Saf.* **2020**, *19*, 3886–3915. [CrossRef] [PubMed]
173. Mairal, T.; Nadal, P.; Svobodova, M.; O'Sullivan, C.K. FRET-based dimeric aptamer probe for selective and sensitive Lup an 1 allergen detection. *Biosens. Bioelectron.* **2014**, *54*, 207–210. [CrossRef] [PubMed]
174. O' Sullivan, C.K.; Mairal, T.; Jauset-Rubio, M.; Svobodova, M.; Skouridou, V.; Esposito, V.; Virgilio, A.; Galeone, A. Aptamers Against the beta-conglutin allergen: Insights into the behavior of the shortest multimeric(Intra)molecular DNA G-Quadruplex. *Int. J. Mol. Sci.* **2021**, *22*, 1150. [CrossRef] [PubMed]
175. Weng, X.; Neethirajan, S. A microfluidic biosensor using graphene oxide and aptamer-functionalized quantum dots for peanut allergen detection. *Biosens. Bioelectron.* **2016**, *85*, 649–656. [CrossRef]

foods MDPI

Communication

Fluorescent and Colorimetric Dual-Mode Strategy Based on Rhodamine 6G Hydrazide for Qualitative and Quantitative Detection of Hg^{2+} in Seafoods

Ziwen Zhang [1,2], Ran Han [1,2], Sixuan Chen [1,2], Feilin Zheng [1,2], Xinmiao Ma [1,2], Mingfei Pan [1,2,*] and Shuo Wang [1,2]

1 Key Laboratory of Food Quality and Health of Tianjin, Tianjin University of Science and Technology, Tianjin 300457, China
2 State Key Laboratory of Food Nutrition and Safety, Tianjin University of Science & Technology, Tianjin 300457, China
* Correspondence: panmf2012@tust.edu.cn; Tel.: +86-22-60912493

Abstract: In this study, a rapid fluorescent and colorimetric dual-mode detection strategy for Hg^{2+} in seafoods was developed based on the cyclic binding of the organic fluorescent dye rhodamine 6G hydrazide (R6GH) to Hg^{2+}. The luminescence properties of the fluorescent R6GH probe in different systems were investigated in detail. Based on the UV and fluorescence spectra, it was determined that the R6GH has good fluorescence intensity in acetonitrile and good selective recognition of Hg^{2+}. Under optimal conditions, the R6GH fluorescent probe showed a good linear response to Hg^{2+} (R^2 = 0.9888) in the range of 0–5 μM with a low detection limit of 2.5×10^{-2} μM (S/N = 3). A paper-based sensing strategy based on fluorescence and colorimetric analysis was developed for the visualization and semiquantitative analysis of Hg^{2+} in seafoods. The LAB values of the paper-based sensor impregnated with the R6GH probe solution showed good linearity (R^2 = 0.9875) with Hg^{2+} concentration in the range of 0–50 μM, which means that the sensing paper can be combined with smart devices to provide reliable and efficient Hg^{2+} detection.

Keywords: Hg^{2+}; R6GH; dual mode; fluorescence; visualization

Citation: Zhang, Z.; Han, R.; Chen, S.; Zheng, F.; Ma, X.; Pan, M.; Wang, S. Fluorescent and Colorimetric Dual-Mode Strategy Based on Rhodamine 6G Hydrazide for Qualitative and Quantitative Detection of Hg^{2+} in Seafoods. *Foods* **2023**, *12*, 1085. https://doi.org/10.3390/foods12051085

Academic Editor: Antonello Santini

Received: 17 January 2023
Revised: 26 February 2023
Accepted: 1 March 2023
Published: 3 March 2023

1. Introduction

In recent years, the rapid development of industry has brought serious pollution to the natural and food production environment. Unlike other types of pollution, heavy metal pollution can circulate within the environment and have the characteristic of being non-degradable, thus causing more serious harm and impact on human beings [1]. Heavy metal ions are capable of denaturing proteins in the living organisms. When these harmful heavy metals accumulate and concentrate in the human body, they can produce an accumulation of toxicity and cause very serious diseases [2,3]. Heavy metal contamination in food has been an important factor affecting food safety. Mercury (Hg) is one of the common heavy metals that can easily cause environmental pollution [4]. Usually, microorganisms in the soil can methylate the Hg element, making it easy to easily be absorbed by microorganisms and thus enter the food chain. By accumulating in the human body for a long time, high concentrations of Hg can produce very serious hazards [5]. Therefore, it is necessary and significant to monitor and detect Hg levels in the environment and food samples.

Currently, analytical strategies such as mass spectrometry and spectroscopy based on large-scale instruments, such as inductively coupled plasma mass spectrometry (ICP-MS) [6–8], atomic absorption spectrometry (AAS) [9,10], and atomic fluorescence spectrometry (AFS) [11,12], are still the main means for accurate detection of heavy metals (including Hg) in various samples, and such methods have unparalleled advantages in terms of detection accuracy and sensitivity. However, such advanced instruments are usually expensive and large, and require relatively complex sample pretreatment processes, which are inadequate for low-cost screening, in situ detection, and large-scale penetration. To avoid these problems, there is an urgent need to establish a fast, sensitive, and portable detection method for heavy metal targets [13,14]. The paper-based sensing strategy based on fluorescent and colorimetric dual mode can achieve naked-eye visualization of the target and portable, low-cost semiquantitative detection [15,16]; on the other hand, this dual-mode spectral signal output can largely guarantee the reliability of the detection results.

Due to the long emission wavelength, low biotoxicity, and pronounced color change, rhodamine-based compounds are commonly used as fluorescent and colorimetric labeling reagents in visualization assays [17–19]. These compounds can specifically complex or bind metal ions or organic small molecules while processing certain recognition abilities. In particular, rhodamine derivatives with spiro ring structure have different optical properties in open and closed loops, making them ideal materials for the construction of optical sensors. It is worth noting that rhodamine 6G hydrazide (R6GH) is one of the most important intermediates of rhodamine compounds, which attracted extensive attention in heavy metal detection studies [20,21]. The amide spiral ring structure with rhodamine as the parent nucleus has an "On–Off" feature. When specific metal ions are added, the amide ring will be opened, resulting in the rupture of the organic dye and enhanced fluorescence [22]. In our previous work [23], the R6GH dye with a spiral ring structure was found to be opened by Pb^{2+}, causing a significant fluorescence signal (Ex: 552 nm). Based on this, a dual-mode fluorescence and colorimetric detection strategy was further designed and constructed for the rapid and efficient detection of Pb^{2+} in water and food samples.

Since heavy metal Hg^{2+} can also trigger the fluorescence switch of the R6GH probe, this study continued to explore the fluorescence response performance of the R6GH probe to the heavy metal Hg^{2+} under different solution systems and thus developed an effective fluorescence analysis strategy for Hg^{2+} (Figure 1). The study further developed portable detection test strips that not only allowed for the naked-eye colorimetric semiquantitative analysis of the Hg^{2+} content but also can be combined with a portable color reader for the rapid screening of target Hg^{2+} in a large number of samples. This study is of great interest for the development of effective strategies for the on-site detection and large-scale screening of trace hazardous substances in food.

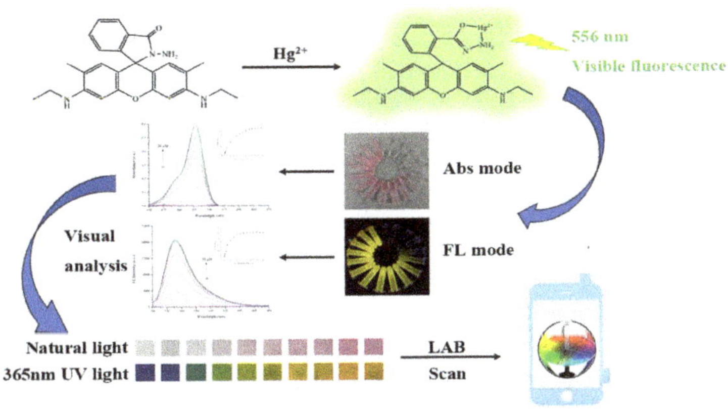

Figure 1. Schematic of R6GH synthesis and fluorescent and colorimetric dual-mode detection of Hg^{2+} for smartphone-integrated sensing system.

2. Materials and Methods

2.1. Reagents and Materials

The reagents and solvent nitrates of target Hg^{2+} and other ions Cu^{2+}, V^{2+}, Cd^{2+}, Mn^{2+}, Zn^{2+}, Cr^{3+}, Co^{2+}, Ag^{+}, and K^{+}; hydrazine hydrate (85%); rhodamine 6G (R6G, 99.5%); and ethylenediaminetetraacetic acid (EDTA) were purchased from Shanghai Aladdin Biochemical Technology Co., Ltd. (Shanghai, China). Organic reagents such as tetrahydrofuran (THF), methanol (MeOH), ethanol (EtOH), and acetonitrile (ACN) were purchased from Sinopharm Chemical Reagent Company. All the reagents used for R6GH probe synthesis and analysis were of analytical grade or higher and were not further purified.

2.2. Instruments

A Shimadzu UV-vis spectrophotometer (Tokyo, Japan, UV-2600) was used to record UV absorbance data, and a Thermo fluorescence spectrometer (Boston, MA, USA, Lumina) was applied for fluorescence analysis. An electric blast dryer (WG II-45BE, Tianjin, China) and a Teflon digestion tank were used for sample pretreatment. An Agilent inductively coupled plasma mass spectrometer (ICP-MS) (Santa Clara, CA, USA, 7700x) was used to compare and validate the data from the Hg^{2+} analysis strategy established in this study. An ordinary quantitative filter paper was used in the visual analysis, and a portable colorimeter from FRU Weifu Optoelectronics (WR-10, Wuxi, China) was used for LAB analysis in the paper-based assays.

2.3. Preparation of R6GH Probe

The R6GH fluorescent probe was prepared according to the method reported [23] and is briefly described as follows: accurately weighed R6G (0.5 g) was fully dissolved in EtOH (30.0 mL) in a 100 mL round-bottom glass flask, and a solution of hydrazine hydrate (85%, 2.0 mL) was added for thorough mixing. The mixture was cooled to room temperature and magnetically stirred for 10 h until the color of the mixed solution disappeared. After filtering under reduced pressure and washing with EtOH three times, the white solid product was collected as the R6GH fluorescent probe.

2.4. Optimization of Dual-Mode Detection System

The prepared R6GH fluorescent probes were dispersed in THF/H_2O (v/v, 1:1), ACN, and MeOH/H_2O (v/v, 3:1), respectively. The concentration of R6GH was controlled at 1.0 mM as a stock solution and diluted to the desired concentration for testing. Subsequently, equal volumes of R6GH solutions at 20 μM concentration in the three systems were mixed with different concentrations (0–60 μM) of Hg^{2+} to obtain the resulting mixtures with Hg^{2+} concentrations ranging from 0 to 30 μM. The mixture was reacted at room temperature for 10 min and then analyzed by UV and fluorescence spectroscopy. By comparing the results, the optimal system for the detection of Hg^{2+} was finally determined and used in the subsequent experiments.

2.5. Detection Procedure for Hg^{2+}

Different concentration gradients (0–10 μM) of Hg^{2+} were added to the ACN system of the R6GH probe (20 μM), thoroughly mixed, and left at room temperature for fluorescence intensity detection. The common metal ions Cu^{2+}, V^{2+}, Cd^{2+}, Mn^{2+}, Zn^{2+}, Cr^{3+}, Co^{2+}, Ag^{+}, and K^{+} (10 μM) in food were added to the R6GH probe solution in the ACN system in the same way, and after sufficient reaction, the fluorescence intensity was compared to assess the interference of different metal ions with Hg^{2+}.

The reproducibility and reversible mechanism of the R6GH probe for the detection of Hg^{2+} were also investigated using ethylenediaminetetraacetic acid (EDTA) to assess the reversibility of R6GH for the detection of Hg^{2+}.

2.6. Preparation of Test Strips for Visual Detection of Heavy Metal Ions

Test strips (1 cm × 1 cm) were infiltrated in the solution of the R6GH probe (20 μM) in the THF/H_2O (v/v, 1:1) solution and left for 10 min at room temperature. The solvent was removed from the test strips and allowed to dry. The paper with the R6GH probe immobilized was cut into strips and fixed, infiltrated in Hg^{2+} standard solutions with different concentration gradients (0–100 μM), and naturally dried to obtain color-developed immobilized paper-based sensors that can be used for naked-eye or fluorescence (365 nm UV) analysis and comparison of Hg^{2+}.

2.7. Actual Sample Pretreatment

Seafood samples of oysters, yellow croaker, and prawn were purchased from a local market in Tianjin and stored in a refrigerator at 4 °C. The samples were first nitrated by adding 0.5 g of the actual sample and 10.0 mL of HNO_3 together at room temperature into a Teflon beaker for an overnight treatment, then boiled until all dissolved, and the cooled solution was centrifuged at 8000 r/min to obtain the supernatant, which was adjusted to pH 6.0 using the NaOH solution (1 mol/L) and diluted to 50 mL with ultrapure water to prepare different concentration gradients of Hg^{2+}.

3. Results and Discussion

3.1. Optimization of Hg^{2+} Assay System

The prepared R6GH probe can react with the target Hg^{2+}, which can open the ring structure it possesses, leading to the enhanced fluorescence. The fluorescence response of the R6GH probe to the target Hg^{2+} in three solutions of THF/H_2O (v/v, 1:1), ACN, and MeOH/H_2O (v/v, 3:1) was investigated, and the sensitivity and accuracy for the dual-mode detection of Hg^{2+} in different systems were compared based on the Hg^{2+}-induced changes in color and fluorescence intensity of the R6GH probe observed under natural and UV light.

Figure 2 has shown the color change of the R6GH probe in the three detection systems under natural light and UV after Hg^{2+} induced the colorimetric and fluorescence switching of the R6GH probe in the "On" state. When Hg^{2+} was added into the solution of the R6GH probe, the solution color changed from colorless to pink under natural light, and the fluorescence in the solution changed from no fluorescence to bright yellow fluorescence under the excitation of 365 nm UV light. Meanwhile, with the increase in Hg^{2+} concentration (0–30 μM), the solution color of the R6GH probe gradually deepened and stabilized. From these results, the color change of R6GH in the THF/H_2O (v/v, 1:1) system was more obvious, with higher chromogen and brighter fluorescence produced, which was easier to observe. Therefore, the THF/H_2O (v/v, 1:1) solution was considered more suitable for the visualization and fluorescent colorimetric analysis of Hg^{2+} by the R6GH probe and used for the detection process of Hg^{2+} colorimetric test strips.

To further explore the Hg^{2+} recognition performance of the R6GH probe in fluorescence and colorimetric detection, the UV absorbance and fluorescence intensity of the R6GH probe were investigated in three systems under the colorimetric and fluorescence "On" states induced by Hg^{2+}, and it was found that the UV absorption and fluorescence intensity significantly increased with the addition of Hg^{2+}. As illustrated in Figure 3, the R6GH probe has no UV absorption and fluorescence emission ability but appeared a UV absorption band near 530 nm and a clear fluorescence emission peak near 556 nm. By comparing the UV absorption and fluorescence spectra in the three detection systems, the ACN system obtained the highest absorbance and fluorescence intensity. This was because the ACN was one nonprotonic solvent that did not provide nor spontaneously transfer protons in the reaction, thus having good solubility for metal cations. Meanwhile, considering the good solubility of ACN for the R6GH probe, it allowed Hg^{2+} to form ion-dipole bonds with the solvent system, thus increasing the contact area of the complexation reaction between the R6GH fluorescent probe and Hg^{2+}, which made the reaction faster and more adequate.

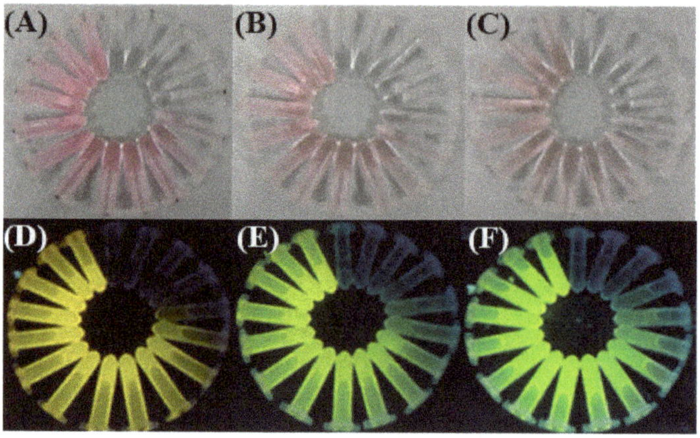

Figure 2. Color changes of Hg^{2+} induced R6GH probes in the three different solutions under natural light ((**A**): THF/H_2O (v/v, 1:1); (**B**): ACN; (**C**): MeOH/H_2O (v/v, 3:1)) and under 365 nm UV light ((**D**): THF/H_2O (v/v, 1:1); (**E**): ACN; (**F**): MeOH/H_2O (v/v, 3:1)).

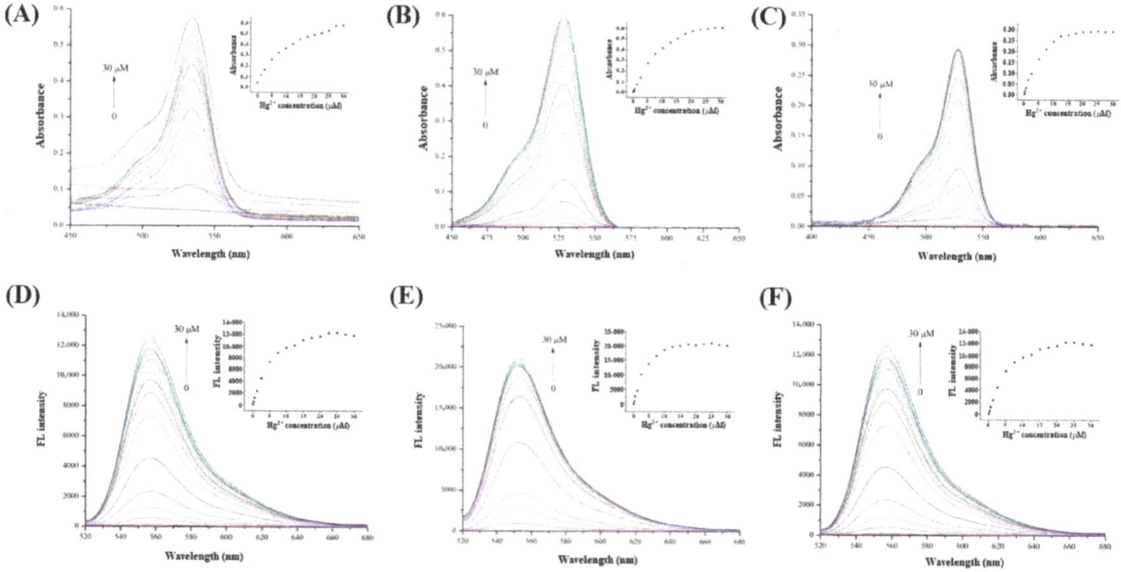

Figure 3. UV absorption spectra ((**A**) THF/H_2O (v/v, 1:1); (**B**) ACN; (**C**) MeOH/H_2O (v/v, 3:1)) and fluorescence spectra ((**D**) THF/H_2O (v/v, 1:1); (**E**) ACN; (**F**) MeOH/H_2O (v/v, 3:1)) of the R6GH probe (10 μM) in three tested solutions complexing with different concentrations of Hg^{2+} (0–30 μM). Inset: Relationship between R6GH absorbance or fluorescence intensity (Em: 556 nm) and Hg^{2+} concentrations.

3.2. Establishment of R6GH-Based Fluorescence Strategy for Hg^{2+} Detection

Hg^{2+} can ligand-complex with the O atom of -COOH and the N atom of -NH_2 in the R6GH probe [24], which induces the conversion of the rhodamine-based amide spiro ring structure from "Off" to "On" state, resulting in the color change or fluorescence enhancement of the originally colorless and nonfluorescent R6GH probe. The fluorescence intensity of the R6GH probe in ACN gradually increased with the increase in Hg^{2+} concentration, having a good linear relationship with Hg^{2+} concentration in the range of 0–5 μM with R^2

of 0.9888 (Figure 4A). The limit of detection (LOD, S/N = 3) of Hg^{2+} reached 2.5×10^{-2} μM, indicating that the prepared R6GH fluorescent probe can sensitively respond to Hg^{2+}.

The selectivity of the R6GH fluorescent probe for Hg^{2+} in the ACN system was further evaluated for common metal ions (Cd^{2+}, Mn^{2+}, V^{2+}, Cu^{2+}, Zn^{2+}, Cr^{3+}, Co^{2+}, Ag^{+}, and K^{+}) in the study. It was found that these metal ions did not significantly enhance the fluorescence intensity of R6GH. When Hg^{2+} (10 μM) induced the fluorescence conversion of the R6GH probe to "open loop", its fluorescence intensity was as high as 9435.1, which indicated that Hg^{2+} had a relatively obvious fluorescence enhancement effect on the R6GH probe. Except for the selected Cd^{2+}, Mn^{2+}, V^{2+}, and Cu^{2+}, which had weak enhancement on the fluorescence of R6GH [23] (approximately 20–50% of the fluorescence intensity of the same concentration of Hg^{2+}), the other tested ions could not produce fluorescence enhancement on the R6GH probe. Furthermore, the interference factor K ($F_{others}/F_{Hg^{2+}}$) was used to compare and assess the interference of other metal ions on the fluorescence response of target Hg^{2+}. Based on the fluorescence intensity of tested metal ions at 5.0 μM, the calculated K values 0.47 (Cd^{2+}), 0.29 (Mn^{2+}), 0.27 (V^{2+}), 0.24 (Cu^{2+}), 0.002 (Ag^{+}), 0.0014 (Zn^{2+}), 0.0012 (K^{+}), 0.00094 (Co^{2+}), and 0.0008 (Cr^{3+}) were all less than 1.0, further demonstrating that the prepared R6GH probe has good selectivity for Hg^{2+}, which provides a great feasibility and theoretical basis for its future practical application.

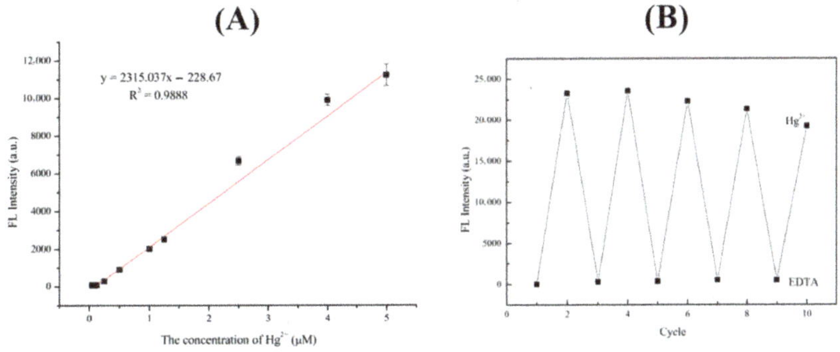

Figure 4. (**A**) Correlation between Hg^{2+} concentrations and R6GH fluorescence intensity. (**B**) Fluorescence reversibility of R6GH under alternate addition of Hg^{2+} and EDTA.

EDTA can coordinate with a variety of metal ions to form complexes and was used to examine the fluorescence reversibility of the R6GH probe [25]. The results showed that the fluorescence intensity of the R6GH–Hg^{2+} system to ACN sharply decreased after the addition of EDTA and even almost restored to the original state of the R6GH probe (Figure 4B). This indicated that EDTA could release Hg^{2+} from the R6GH–Hg^{2+} complex and turn off the fluorescence switch of the R6GH probe. When Hg^{2+} was added again, the fluorescence intensity of the solution recovered to be close to that of the R6GH probe solution when only Hg^{2+} was present, indicating that EDTA can cause the demetallization of the R6GH probe and regeneration of the spirolactam ring. After five cycles, the fluorescence of the R6GH probe solution did not significantly increase or decrease, indicating that the R6GH probe can release Hg^{2+} through competition with EDTA for reversible cycling of Hg^{2+} detection.

3.3. Detection of Hg^{2+} in Seafoods Using R6GH-Based Fluorescent Probes

To verify the application capability of the prepared R6GH-based fluorescent probe, seafoods including oysters, yellow croaker, and prawn were selected and spiked Hg^{2+} with different concentrations (0.5, 2.0, and 4.0 μM) to perform the recovery experiments. The spiked samples were simply pretreated and used for the constructed R6GH-based fluorescent probes and widely accepted ICP-MS methods for detection. As shown in Table 1, the proposed R6GH-based fluorescent probe obtained acceptable recoveries (88.0–108.3%)

of Hg^{2+} in each selected seafood with RSDs all below 5% (n = 3). Compared with the results obtained from the conventional ICP-MS method, good correlation was achieved with $r^2 > 0.99$ (Figure 5). These results indicated that the proposed R6GH probe-based fluorescence strategy was an ideal tool for providing accurate and reliable detection of Hg^{2+} in food matrices. Table 2 has compared the merits of different strategies for Hg^{2+} detection, signifying that the proposed R6GH-based probe can offer a rapid, sensitive, and effective strategy for Hg^{2+}.

Table 1. Results of Hg^{2+} detection in seafood samples using R6GH probe.

Seafood	Spiked Level (μM)	R6GH-Based Fluorescence Strategy			ICP-MS Method		
		Found (μM)	Recovery (%)	RSD (%, n = 3)	Found (μM)	Recovery (%)	RSD (%, n = 3)
Oysters	0.5	0.45	89.2	4.7	0.46	92.0	3.6
	2.0	1.76	88.0	3.5	1.88	94.0	3.2
	4.0	3.65	91.3	3.2	3.83	95.8	2.3
Yellow croaker	0.5	0.49	97.2	4.4	0.47	94.0	4.1
	2.0	1.89	94.7	2.8	1.94	97.0	3.1
	4.0	4.33	108.3	3.3	4.10	102.5	2.5
Prawn	0.5	0.52	103.6	4.0	0.50	100.0	3.7
	2.0	1.88	94.0	3.8	1.96	98.0	2.2
	4.0	3.75	93.8	2.4	3.82	95.5	1.9

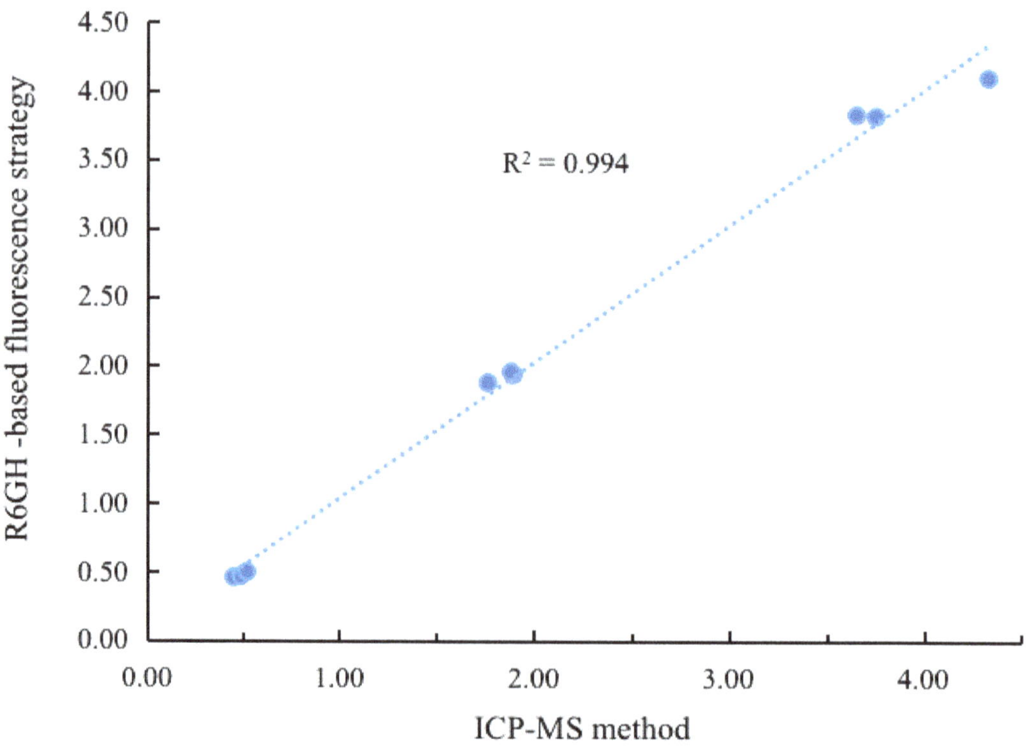

Figure 5. Correlation of Hg2+ detection results in real samples by R6GH-based fluorescence strategy and ICP-MS method.

Table 2. Comparison of merits of different Hg^{2+} detection methods.

Methods	Materials	Linear Range	LOD	Required Time	Ref.
Multicapillary GC-ICP-MS	-	0.002–10 pg mL^{-1}	0.08 pg	-	[6]
ICP-MS/MS	-	1.7–325.6 ng g^{-1}	0.85 ng L^{-1}	-	[7]
Fluorescent	Carbon nanodots	0–3 μM	4.2 nM	~10 min	[26]
Ultraviolet spectrophotometry	Gold Nanorods	285 nM–8.00 μM	112 nM	-	[27]
Ratiometric fluorescent paper	Dual-colored carbon dots	0–320 nM	0.14 nM	~3 min	[28]
Electrochemical biosensor	Poly-T oligonucleotides	1 nM–1.0 mM	100 pM	~30 min	[29]
Fluorimetry and visualization assay	R6GH	0–5 μM	0.025 μM	<10 min	This work

3.4. Development of Visualization Paper-Based Sensor for Hg^{2+} Detection

Fluorescence sensing analysis suffers from the shortcoming that single signal readings are susceptible to environmental and human factors; thus, there is an urgent need to develop new sensing platforms capable of rapid, on-site, and reliable heavy metal ion detection and analysis. Based on the constructed fluorescence sensing system of R6GH–Hg^{2+}, an intelligent, low-cost, and portable paper-based sensing platform was developed to realize fluorescent and colorimetric dual-mode signal output for more accurate, reliable, and convenient detection of heavy metal Hg^{2+}. As shown in Figure 6A, the paper-based sensor constructed with a filter paper infiltrated with the R6GH–Hg^{2+} solution as a substrate (the paper used in the experiment has blue background fluorescence) showed a colorless to light pink change (natural light) and a blue to yellow-green change (UV light, 365 nm) with increasing Hg^{2+} concentration from 0 to 100 μM.

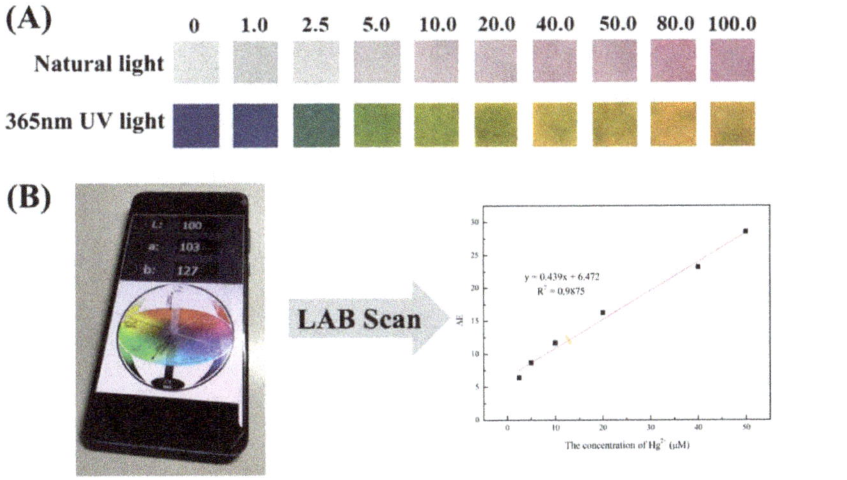

Figure 6. (**A**) Colors of immobilized paper-based sensors containing R6GH immersed in different concentrations of Hg^{2+} under natural and UV light. (**B**) Calibration curve of LAB ΔE value to Hg^{2+} concentration.

For visualization of the results, colorimetric signals can be recorded and analyzed using a colorimeter to provide the LAB values of the test strips. The LAB color model consists of three elements, luminance L and the associated colors A and B [30,31]. The LAB color space defining the color change can be further linked to digital cameras and smartphones, thus facilitating remote monitoring and online analysis. Thus, the chromatic aberration parameter ΔE can be used as a reference for the visual detection of Hg^{2+} by paper-based sensors. The chromaticity difference at different concentrations of Hg^{2+} can be calculated according to the equation ($\Delta E = \sqrt{\Delta L^2 + \Delta A^2 + \Delta B^2}$) [32]. The results showed a good linear relationship ($y = 0.439x + 6.472$, $R^2 = 0.9875$) between the chromatic aberration parameter ΔE and Hg^{2+} concentration for the paper-based sensor infiltrated using the

R6GH–acetonitrile probe solution in the Hg^{2+} concentration range of 2.5–50 μM (Figure 6B). Therefore, the paper-based array sensor constructed based on fluorescence and colorimetric detection strategies combined with a smart device can achieve visualized semiquantitative detection of target Hg^{2+}. This dual-mode sensing method based on R6GH enabled more convenient, reliable, and accurate analysis of target Hg^{2+}.

4. Conclusions

In summary, this study successfully synthesized the R6GH probe that generates fluorescence and color signals with Hg^{2+} and developed a fluorescence-colorimetric dual-mode sensing platform that can be used for rapid, accurate, and sensitive detection of Hg^{2+} in seafood. This R6GH probe was used not only to construct a fluorescence analysis platform with good linearity, accuracy, and sensitivity for Hg^{2+} but also to develop a paper-based visual semiquantitative analysis strategy that can work in conjunction with a small and portable color reader, providing an ideal tool for rapid, low-cost, and convenient analysis of Hg^{2+}. The R6GH fluorescent probe-based strategy and research method proposed in this study can be extended to the detection of other targets in other fields, providing a new direction for the research of high-performance and intelligent analytical strategies and detection devices.

Author Contributions: Conceptualization, Z.Z. and M.P.; methodology, Z.Z.; validation, Z.Z. and R.H.; formal analysis, S.C.; data curation, Z.Z. and F.Z.; writing—original draft preparation, Z.Z. and X.M.; writing—review and editing, M.P. and S.W.; supervision, S.W.; project administration, M.P.; funding acquisition, M.P. All authors have read and agreed to the published version of the manuscript.

Funding: This work was supported by the National Natural Science Foundation of China (No. 31972147 and 32272416), Project of Tianjin Science and Technology Plan (No. 22ZYJDSS00030), and Project Program of Key Laboratory of Tianjin Key Laboratory of Food Quality and Health, China (No. TJS202205).

Institutional Review Board Statement: Not applicable.

Informed Consent Statement: Not applicable.

Data Availability Statement: Not applicable.

Conflicts of Interest: The authors declare no conflict of interest.

References

1. Ng'ang'a, J.; Fombong, F.; Kiiru, S.; Kipkoech, C.; Kinyuru, J. Food safety concerns in edible grasshoppers: A review of microbiological and heavy metal hazards. *Int. J. Trop. Insect Sci.* **2021**, *41*, 2103–2111. [CrossRef]
2. Nduka, J.K.; Orisakwe, O.E. Heavy metal hazards of pediatric syrup administration in Nigeria: A look at chromium, nickel and manganese. *Int. J. Environ. Res. Public Health* **2009**, *6*, 1972–1979. [CrossRef] [PubMed]
3. Awad, M.; El-Sayed, M.M.; Li, X.; Liu, Z.; Mustafa, S.K.; Ditta, A.; Hessini, K. Diminishing heavy metal hazards of contaminated soil via biochar supplementation. *Sustainability* **2021**, *13*, 12742. [CrossRef]
4. Suren, S.; Punyain, W.; Maneeintr, K.; Nootong, K.; Pancharoen, U. The simultaneous elimination of arsenic and mercury ions via hollow fiber supported liquid membrane and their reaction mechanisms: Experimental and modeling based on DFT and generating function. *Arabian J. Chem.* **2022**, *16*, 104501. [CrossRef]
5. Liu, G.; Cai, Y.; O'Driscoll, N.J. Environmental chemistry and toxicology of mercury. *Ecotoxicol. Environ. Saf.* **2020**, *202*, 110926.
6. Pereiro, I.R. Optimization of the coupling of multicapillary GC with ICP-MS for mercury speciation analysis in biological materials. *J. Anal. At. Spectrom.* **1999**, *14*, 851–857.
7. Fu, L.; Shi, S.Y.; Chen, X.Q. Accurate quantification of toxic elements in medicine food homologous plants using ICP-MS/MS. *Food Chem.* **2018**, *245*, 692–697. [CrossRef]
8. Pérez-Álvarez, E.P.; García, R.; Barrulas, P.; Dias, C.; Cabrita, M.J.; Garde-Cerdán, T. Classification of wines according to several factors by ICP-MS multi-element analysis. *Food Chem.* **2019**, *270*, 273–280. [CrossRef]
9. Domanico, F.; Forte, G.; Majorani, C.; Senofonte, O.; Petrucci, F.; Pezzi, V.; Alimonti, A. Determination of mercury in hair: Comparison between gold amalgamation-atomic absorption spectrometry and mass spectrometry. *J. Trace Elem. Med. Biol.* **2017**, *43*, 3–8. [CrossRef]
10. Frois, C.F.; Boschetti, W.; dos Passos, A.S.; Potes, M.L.; Vale, M.G.R.; Silva, M.M. A comparison between chemical and photochemical vapor generation techniques for mercury determination using univariate and multivariate optimization. *Microchem. J.* **2020**, *157*, 105029. [CrossRef]

11. Grijalba, A.C.; Quintas, P.Y.; Fiorentini, E.F.; Wuilloud, R.G. Usefulness of ionic liquids as mobile phase modifiers in HPLC-CV-AFS for mercury speciation analysis in food. *J. Anal. At. Spectrom.* **2018**, *33*, 822–834. [CrossRef]
12. da Silva, D.L.F.; da Costa, M.A.P.; Silva, L.O.B.; Dos Santos, W.N.L. Simultaneous determination of mercury and selenium in fish by CVG AFS. *Food Chem.* **2019**, *273*, 24–30. [CrossRef] [PubMed]
13. Pak, K.R.; Bartha, R. Mercury methylation and demethylation in anoxic lake sediments and by strictly anaerobic bacteria. *Appl. Environ. Microbiol.* **1998**, *64*, 1013–1017. [CrossRef]
14. Pundi, A.; Chang, C.J. Recent advances in synthesis, modification, characterization, and applications of carbon dots. *Polymers* **2022**, *14*, 2153. [CrossRef] [PubMed]
15. Zhang, X.; Deng, J.; Xue, Y.; Shi, G.; Zhou, T. Stimulus response of Au-NPs@ GMP-Tb core–shell nanoparticles: Toward colorimetric and fluorescent dual-mode sensing of alkaline phosphatase activity in algal blooms of a freshwater lake. *Environ. Sci. Technol.* **2016**, *50*, 847–855. [CrossRef]
16. Azmi, A.A.; Daud, A.I.; Khairul, W.M.; Hamzah, S.; WMA, W.M.K.; Hairom, N.H.H. Silica–silver core-shell nanoparticles incorporated with cellulose filter paper as an effective colorimetric probe for mercury ion detection in aqueous media: Experimental and computational evaluations. *Environ. Nanotechnol. Monit. Manag.* **2022**, *19*, 100762. [CrossRef]
17. Li, H.; Chen, Q.; Hassan, M.M.; Ouyang, Q.; Jiao, T.; Xu, Y.; Chen, M. AuNS@ Ag core-shell nanocubes grafted with rhodamine for concurrent metal-enhanced fluorescence and surfaced enhanced Raman determination of mercury ions. *Anal. Chim. Acta* **2018**, *1018*, 94–103. [CrossRef]
18. Zhang, Q.; Liu, X.J.; He, R.C.; Guo, C.B.; Zhao, W.Z.; Zeng, C.C.; Yin, L.P. Development of a fluorescent-type sensor based on rhodamine B for Fe (III) determination. *Chem. Lett.* **2018**, *47*, 122–125. [CrossRef]
19. Zhao, M.; Guo, Y.S.; Fu, G.D.; Wang, Q.; Sheng, W.L.; Guo, D.S. Development of a Si-rhodamine-based NIR fluorescence probe for highly specific and quick response of Hg^{2+} and its applications to biological imaging. *Microchem. J.* **2021**, *171*, 106855. [CrossRef]
20. Wang, S.; Li, C.; Xu, L.; Xu, T.; Lv, Y.; Son, Y.A.; Cao, D. A photochromic fluorescent probe for Hg^{2+} based on dithienylethene-rhodamine b dyad and its application in live cells imaging. *Sci. Adv. Mater.* **2017**, *9*, 533–540. [CrossRef]
21. Meng, X.; Li, Z.; Ma, W. A highly sensitivity fluorescent probe based on rhodamine for naked-eye detection of Hg^{2+} in aqueous solution. *Int. J. Environ. Anal. Chem.* **2021**, 1–11. [CrossRef]
22. Rajasekar, M. Recent trends in Rhodamine derivatives as fluorescent probes for biomaterial applications. *J. Mol. Struct.* **2021**, *1235*, 130232. [CrossRef]
23. Xie, X.; Pan, M.; Hong, L.; Liu, K.; Yang, J.; Wang, S.; Wang, S. An "off–on" rhodamine 6G hydrazide-based output platform for fluorescence and visual dual-mode detection of lead (II). *J. Agric. Food Chem.* **2021**, *69*, 7209–7217. [CrossRef] [PubMed]
24. Xu, Y.; Niu, X.; Zhang, H.; Xu, L.; Zhao, S.; Chen, H.; Chen, X. Switch-on fluorescence sensing of glutathione in food samples based on a graphitic carbon nitride quantum dot (g-CNQD)–Hg^{2+} chemosensor. *J. Agric. Food Chem.* **2015**, *63*, 1747–1755. [CrossRef] [PubMed]
25. Lu, B.; Xia, D.; Zhao, S.; Yang, Z.; Chen, Z.; Huang, S.; Zhang, Z. The influence mechanism of ethylenediaminetetraacetic acid (EDTA) and ferrous iron on the bioavailability of Fe in the process of low rank coal fermentation. *Biochem. Eng. J.* **2022**, *185*, 108520. [CrossRef]
26. Zhou, L.; Lin, Y.; Huang, Z.; Ren, J.; Qu, X. Carbon nanodots as fluorescence probes for rapid, sensitive, and label-free detection of Hg^{2+} and biothiols in complex matrices. *Chem. Commun.* **2012**, *48*, 1147–1149. [CrossRef]
27. Duan, Q.Q.; Zhou, J.L.; Li, P.W.; Sun, L.; Zhuo, K.; Zhang, Y.X.; Sang, S.B. High-sensitivity mercury ion detection system using unmodified gold nanorods. *Chinese. J. Anal. Chem.* **2018**, *46*, e1874–e1879. [CrossRef]
28. Wang, Y.; Yang, L.; Liu, B.; Yu, S.; Jiang, C. A colorimetric paper sensor for visual detection of mercury ions constructed with dual-emission carbon dots. *New J. Chem.* **2018**, *42*, 15671–15677. [CrossRef]
29. Cao, R.G.; Zhu, B.; Li, J.; Xu, D. Oligonucleotides-based biosensors with high sensitivity and selectivity for mercury using electrochemical impedance spectroscopy. *Electrochem. Commun.* **2009**, *11*, 1815–1818. [CrossRef]
30. Kim, H.; Kim, H.; Choi, J.; Inn, K.S.; Seong, J. Visualization of autophagy progression by a red–green–blue autophagy sensor. *ACS Sens.* **2020**, *5*, 3850–3861. [CrossRef]
31. Ulrici, A.; Foca, G.; Ielo, M.C.; Volpelli, L.A.; Fiego, D.P.L. Automated identification and visualization of food defects using RGB imaging: Application to the detection of red skin defect of raw hams. *Innov. Food Sci. Emerg. Technol.* **2012**, *16*, 417–426. [CrossRef]
32. Wee, A.G.; Lindsey, D.T.; Kuo, S.; Johnston, W.M. Color accuracy of commercial digital cameras for use in dentistry. *Dent. Mater.* **2006**, *22*, 553–559. [CrossRef] [PubMed]

foods

MDPI

Article

Glutathione-Capped CdTe Quantum Dots Based Sensors for Detection of H_2O_2 and Enrofloxacin in Foods Samples

Shijie Li [1,†], Linqing Nie [1,†], Lin Han [2], Wenjun Wen [3], Junping Wang [3,*] and Shuo Wang [1,*]

[1] Tianjin Key Laboratory of Food Science and Health, School of Medicine, Nankai University, Tianjin 300071, China
[2] Beijing Advanced Innovation Center for Soft Matter Science and Engineering, Beijing University of Chemical Technology, Beijing 100029, China
[3] State Key Laboratory for Food Nutrition and Safety, Tianjin University of Science and Technology, Tianjin 300457, China
* Correspondence: wangjp@tust.edu.cn (J.W.); wangshuo@nankai.edu.cn (S.W.); Tel.: +86-22-85358445 (S.W.)
† These authors contributed equally to this work.

Abstract: Additives and antibiotic abuse during food production and processing are among the key factors affecting food safety. The efficient and rapid detection of hazardous substances in food is of crucial relevance to ensure food safety. In this study, a water-soluble quantum dot with glutathione as a ligand was synthesized as a fluorescent probe by hydrothermal method to achieve the detection and analysis of H_2O_2. The detection limits were 0.61 μM in water and 68 μM in milk. Meanwhile, it was used as a fluorescent donor probe and manganese dioxide nanosheets were used as a fluorescent acceptor probe in combination with an immunoassay platform to achieve the rapid detection and analysis of enrofloxacin (ENR) in a variety of foods with detection limits of 0.05–0.25 ng/mL in foods. The proposed systems provided new ideas for the construction of fluorescence sensors with high sensitivity.

Keywords: manganese dioxide nanosheet; water-soluble quantum dots with glutathione as ligand; fluorescence quenching immunosensors; antibiotic detection

Citation: Li, S.; Nie, L.; Han, L.; Wen, W.; Wang, J.; Wang, S. Glutathione-Capped CdTe Quantum Dots Based Sensors for Detection of H_2O_2 and Enrofloxacin in Foods Samples. *Foods* **2023**, *12*, 62. https://doi.org/10.3390/foods12010062

Academic Editor: Antonello Santini

Received: 2 September 2022
Revised: 8 December 2022
Accepted: 12 December 2022
Published: 22 December 2022

1. Introduction

Food safety is directly related to people's livelihoods, and it is also necessary to promote economic development and social harmony [1]. However, the globalization process of food trade has increased the risk of spreading contaminated food [2]. Food safety, as a global issue, has attracted increasing attention from governments, food industries and consumers. Food safety detection technologies play crucial roles in ensuring the health and safety of food for the population [3]. However, since most contaminants often exist in trace amounts, and complex food matrics seriously interfere with the detection results, the development of detection and analysis methods with higher sensitivity and accuracy has been the pursuit of food safety analysis [4]. For example, veterinary antibiotics such as enrofloxacin (ENR) are not completely biodegradable by animals, and their unfounded use can endanger human health in the form of prototypes or metabolites through the food chain [5–7]. In addition, H_2O_2 is fraudulently used to block microbial activity in milk that is near its sell-by date or unfit for consumption. As well, most countries have established maximum residue limits for veterinary drugs and hydrogen peroxide in food due to their toxic effects [8,9]. Therefore, it is essential to develop rapid and sensitive detection strategies for veterinary antibiotics and H_2O_2.

In the continuous innovation process of nanomaterials science, the detection technology based on fluorescent nanomaterials has gradually replaced the traditional detection and analysis methods based on large instruments due to the advantages of small instrument dependence, high signal sensitivity and short detection time, and has become a new direction for the development of food safety detection technology. The introduction of semiconductor quantum dots (QDs), up-conversion nanoparticles (UCNPs), atomic clusters (NCs), carbon dots (CDs) and other new fluorescent nanomaterials have promoted the development of fluorescent labeling technology, broadened the application range of fluorescent detection sensors [10–13]. QDs stand out among many fluorescent materials [14,15]. According to the different constituent elements and particle size, QDs achieve full coverage of the emission spectrum from the visible spectrum to the mid-infrared region [16,17]. Especially water-soluble QDs with biomolecules as ligands, due to their good biocompatibility, high quantum yield, and stable optical properties, have become a better choice for fluorescently labeled probes, and have a wide range of applications in fields of biochemistry, immunosensing, cell biology researches and so on [18,19]. Water-soluble QDs with GSH as a ligand (GSH-QDs) have a high degree of biocompatibility, mono-dispersity and stabilities [20–22]. The environmentally friendly GSH-QDs are easy to synthesize and easy to realize industrial production, which has promising commercial application prospects. In addition to its strong coordination ability, GSH also has a strong reducing ability, which can reduce MnO_2 to Mn^{2+}, and the -SH is oxidized to -S-S- to form GSSG [23]. Based on this, the redox reaction system between GSH and MnO_2 is widely used in the field of biological monitoring, however, most reports still focused on the application of MnO_2-based glutathione detection, and its contribution in the field of other target sensing analysis remains to be developed [24–26].

Nanosheets are a novel class of nanomaterials, which have been widely used in the fields of nanotechnology and nanomaterials [27]. Layered manganese dioxide nanosheets (MnO_2 NSs) with high degrees of freedom are an important 2D layered functional material, which has attracted much attention due to its excellent optical properties and adsorption properties while retaining oxidative properties [28]. Based on the unique characteristics of GSH, water-soluble CdTe QDs with GSH as the ligand (GSH-CdTe QDs) were synthesized by hydrothermal method as the fluorescence signal probe. Firstly, an H_2O_2 detection method based on GSH-CdTe QDs was established for the highly sensitive detection of H_2O_2 in milk. In addition, a novel MnO_2 NSs/GSH-CDTE QDs fluorescence quenching sensing system was constructed by coupling the detection antibody to the surface of MnO_2 NSs as a sensing probe. Combined with the immunochromatographic analysis platform, a novel fluorescence quench immunosensor (FQISs) with strong anti-interference ability and high sensing sensitivity were constructed for the detection of ENR, and it is expected to be used for the detection and analysis of more trace hazards in food.

2. Materials and Methods

2.1. Chemicals and Materials

GSH, cadmium chloride hydrate ($CdCl_2$), tellurium powder, sodium borohydride ($NaBH_4$), NaOH, Bovine serum albumin (BSA), and ovalbumin (OVA), 1-(3-Dimethylamino-propyl)-3-ethylcarbodiimide hydrochloride (EDC), HEPES, NC membrane (Millipore HF90s) were purchased from Merck KGaA, (Darmstadt, Germany). MnO_2 NSs were purchased from Nanjing XFNANO Materials Tech Co., Ltd. (Nanjing, China). PVC sheet, sample pad, conjugate pad and absorption pad were purchased from Shanghai Kinbio Tech.Co.,ltd (Shanghai, China), Enrofloxacin, flumequine, danofloxacin, sparfloxacin, gatifloxacin, fleroxacin, lomefloxacin, difloxacin, sarafloxacin, ciprofloxacin, and norfloxacin were purchased from Dr. Ehrenstorfer GmbH (Augsburg, Germany). A commercial ENR ELISA test kit was purchased from Reagen LLC (Moorestown, NJ, USA). Anti-ENR polyclonal antibody (Ab) was produced in our laboratory.

2.2. Synthesis of GSH-CdTe QDs

GSH-CdTe QDs with a final molar ratio of $Cd^{2+}/Te^{2-}/GSH = 1:0.5:2.5$ was synthesized by following a previous method with a minor modification [29]. A total of 0.25 mmol of Te powder and 1.32 mmol of $NaBH_4$ were added into 3 mL of ultrapure water and stirred for 4 h under a nitrogen atmosphere in an ice bath to form the NaHTe precursor. At the same time, 0.5 mmol of $CdCl_2 \cdot 5H_2O$ and 1.25 mmol of GSH were dissolved into 120 mL of ultrapure water and adjusting the mixture to pH 10 with 1.0 M NaOH to form the cadmium precursor and kept stirred under bubbling nitrogen for 30 min. Then, the freshly NaHTe was added into the cadmium precursor immediately under a nitrogen atmosphere to form the CdTe QDs precursor. After refluxing for 2 h, the CdTe QDs with the maximum emission wavelength of 610 nm were prepared. Finally, the product was purified by centrifugation to remove free GSH and dispersed in ultrapure water and stored at 4 °C for use (Scheme 1).

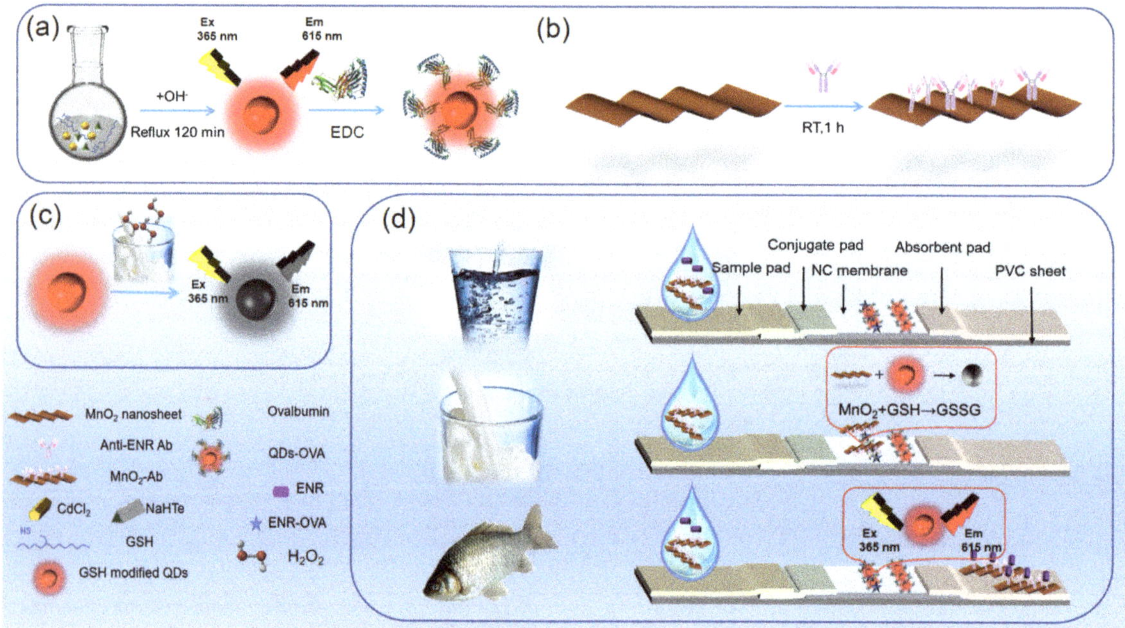

Scheme 1. Schematic of FQISs. Preparation method of the GSH-CdTe QDs-OVA fluorescence probe (**a**) and MnO_2-Ab fluorescence sensing probe (**b**), Detection procedures of FQISs (**c**) and schematic of fluorescence results under a UV lamp (**d**).

2.3. *Fluorescence Responses to H_2O_2 in Milk*

1 mL of milk is dissolved in 9 mL of water, and without centrifugation and other pretreatments, 10 μL of the solution was mixed with 10 μL of GSH-CdTe QDs, diluted with water to 100 μL, and the fluorescence intensity of the solution was measured after standing for 10 min.

2.4. *Fluorescence Responses to MnO_2 NSs*

100 μL of MnO_2 NSs with different concentrations and 100 μL of 200 μg/mL CdTe QDs were added to each microwell in order, and the fluorescence value was immediately measured with a multifunctional fluorescent microplate reader and the fluorescent quenching rate (FQR) was calculated. FQR = $(F_0 - Fx)/F_0 * 100\%$, where F0 is the fluorescence value of QDs without adding MnO_2 NSs, and Fx is the fluorescence value of QDs with MnO_2 NSs added. In order to verify the cause of the fluorescence quenching phenomenon, the above CdTe QDs were replaced with ZnCdSe/ZnS QDs, CDs and RhB and the above experiment were repeated.

2.5. Preparation of Fluorescent Signal Probe (QDs-OVA)

OVA was coupled to the surface of GSH-CdTe QDs by chemical bonding to prepare a fluorescent signal probe. The detailed procedure was as follows: 10 μL of EDC (5 mg) and 10 μL of OVA (10 mg/mL) were added to 200 μL of GSH-CdTe QDs (200 μg/mL) with shaking incubation for 3 h at room temperature, then 10 μL of BSA (200 mg/mL) was added into the mixture and shaken for 10 min. The product was centrifuged to remove agglomerates, and the unconjugated protein was removed by 30 KDa ultrafiltration tubes, the conjugate was re-dissolved in 200 μL PBS and stored at 4 °C.

2.6. Preparation of Fluorescent Quenching Probe (MnO_2 NSs-Ab)

The detection antibody of ENR (ENR-Ab) was coupled to MnO_2 NSs by physical adsorption to prepare a sensor probe. A total of 1 mL of MnO_2 NSs (50 μg/mL) was mixed with 2 μL of K_2CO_3 (0.2 mol/L) solution and 3 μL of ENR-Ab (0.65 mg/mL), and incubated for 1 h at room temperature. After that, 20 μL of 20% BSA solution and 10 μL of 10% $PEG_{20,000}$ was added into the mixture and incubated for 30 min at room temperature, then the mixture was centrifuged at 12,000 rpm for 15 min and the precipitation was resolved into 200 μL of working solution.

2.7. Preparation of Enrofloxacin Coating Antigen (ENR-OVA)

ENR-OVA with a molar ratio of OVA to ENR of 1:25 was synthetized according to the mixed acid anhydride method [30] with slight modification; full details can be found in Supplementary Information (SI).

2.8. Detection Procedure

The details of the preparation of FQISs are provided in the Supplementary Information (SI). After preparing, 100 μL of standard or sample solution and 10 μL of the MnO_2-Ab fluorescent quenching probe were mixed and dropped onto the sample pad with visual results being obtained within 10 min under a UV lamp. The detection principle is also provided in Supplementary Information (SI).

3. Results and Discussion

3.1. Characterization of GSH-CdTe QDs

The synthesis procedure of GSH-CdTe QDs was shown in Scheme 1a, and transmission electron microscopy (TEM), high-resolution TEM (HR-TEM), UV-Vis absorption and fluorescence spectroscopy were used for the characterization of GSH-CdTe QDs. GSH-CdTe QDs with an average diameter of 3.2 nm (Figure S1) had a good dispersive crystal structure, and it was found that the lattice fringe (d = 0.23 nm) corresponded to the (200) crystal plane of CdTe blende (Figure 1a,b), which was consistent with the description in the previous report [31]. In addition, the optical properties of CdTe QDs were verified by fluorescence spectroscopy (Figure 1c,d). GSH-CdTe QDs had a concentration-dependent fluorescence emission peak at 615 nm under excitation at 365 nm, and its fluorescence intensity showed a steady linear decrease trend as the concentration of QDs decreased. The above results demonstrated the successful synthesis of QDs and their feasibility as signal components.

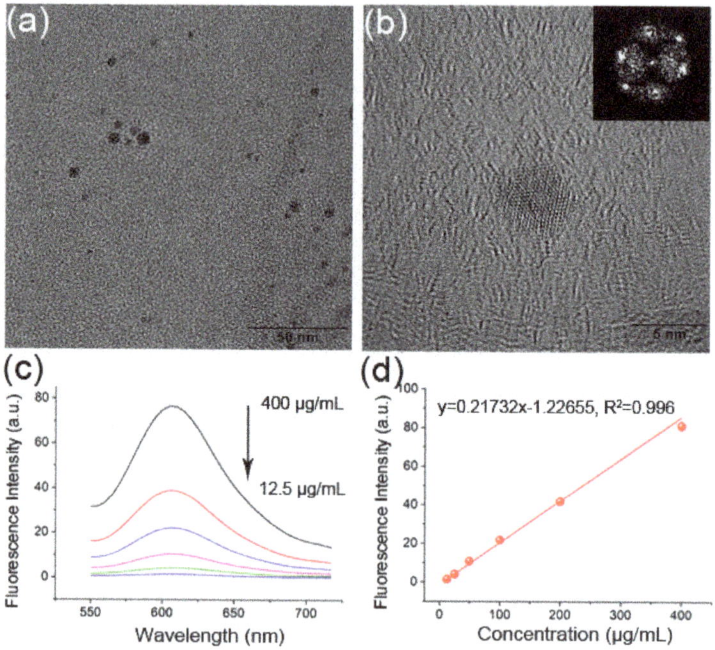

Figure 1. Characterization of GSH-CdTe QDs. TEM images (**a**) and HR-TEM image (Insert: Fourier diffractogram image) (**b**) of GSH-CdTe QDs. Fluorescence spectroscopy (**c**) and the linear relationship between emission fluorescence intensity at 615 nm (**d**) versus 12.5–400 μg/mL of GSH-CdTe QDs.

3.2. *H_2O_2 Detection in Milk Powder*

The fluorescence of GSH-CdTe QDs was quenched in H_2O_2 (Figure S2), which was speculated to be caused by the oxidation of the QDs by the strongly oxidizing H_2O_2 leading to the generation of new defects on their surface, thus increasing the non-radiative transition, which in turn reduced the generation of fluorescence in the excitonic state, leading to fluorescence quenching [8]. It can be seen from Figure 2a that the fluorescence intensity gradually decreased with the increase of H_2O_2 concentration in a logarithmic trend rather than a linear trend, indicating that the quenching of the fluorescence of QDs was achieved through complex multivariate interactions. For the detection of H_2O_2 in water, the linear range was 1.23–300 μM of H_2O_2 with an LOD of 0.61 μM (S/N = 3). H_2O_2 is often used as a bactericide and other food processing aids in the production of milk and dairy products. However, milk has obvious fluorescence in the visible light range due to its complex composition. To verify the feasibility of GSH-CdTe QDs to detect H_2O_2 in milk, the interference of milk samples on QDs fluorescence was first evaluated (Figure 2b). The fluorescence intensity of milk decreased gradually in the range of 450–650 nm and was negligible at 615 nm, the maximum emission wavelength of GSH-CdTe QDs. Therefore, milk hardly interferes with the GSH-CdTe QDs fluorescence signal. In addition, the selectivity of the GSH-CdTe QDs was evaluated to exclude possible interferences in milk samples (Figure S3). The fluorescence values of GSH-CdTe QDs were reduced when the content of κ-casein and BSA was at 1%, but considering the large dilution factor of milk and milk powder in the assay, proteins such as κ-casein can not cause significant effects. Other than that, other interferents did not cause a decrease in the fluorescence of GSH-CdTe QDs. Therefore, the probe was considered suitable for the detection of H_2O_2 in milk. Furthermore, the checkerboard assay was used to evaluate the optimal operating conditions of GSH-CdTe QDs for the detection of H_2O_2 in milk (n = 3) (Figure 2c). The results showed that 1% milk solution containing 0.8 μg/mL GSH-CdTe QDs had the best detection efficiency and sensitivity. Under the above working conditions, the detection method of H_2O_2 in milk was constructed (Figure 2d), and the linear range was 1.23–300 μM of H_2O_2 with the LOD of 0.68 μM (S/N = 3).

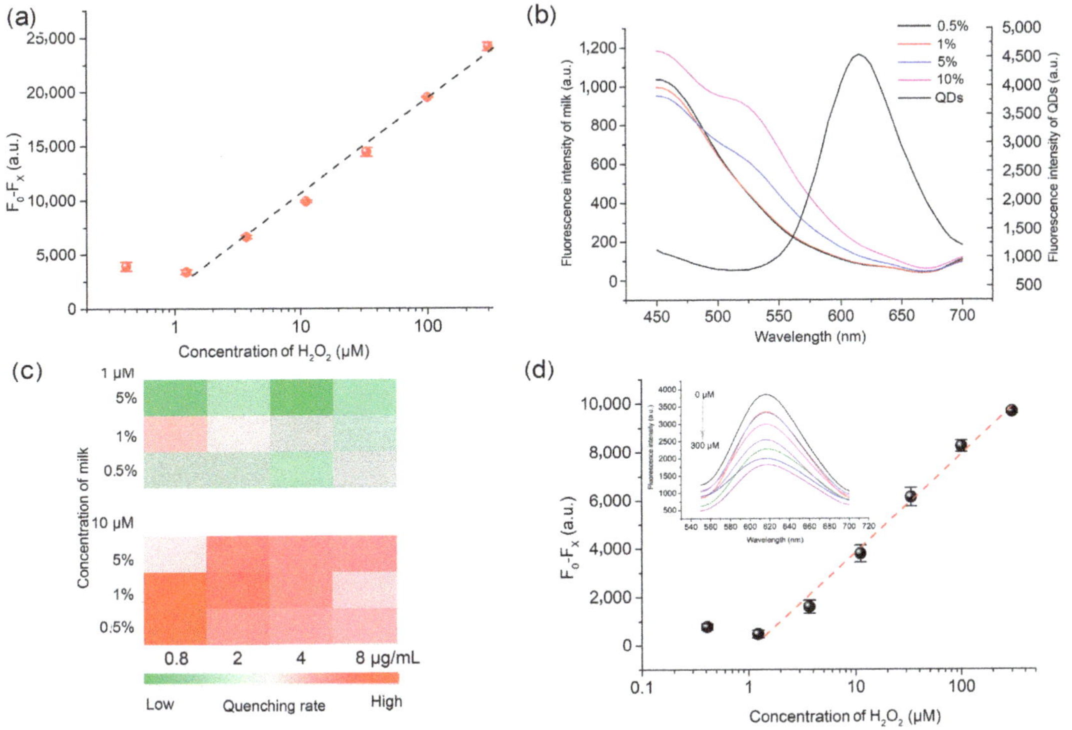

Figure 2. Detection of H_2O_2 by GSH-CdTe QDs in milk. Calibration curve for detection H_2O_2 in water (**a**); Fluorescence spectra of milk and GSH-CdTe QDs (**b**); Optimization results of milk and GSH-CdTe QDs concentration (**c**); And linear standard curve for H_2O_2 detection in Milk (Insert: Fluorescence spectra of GSH-CdTe QDs at each H_2O_2 concentration) (**d**).

Many colorimetric, fluorescent and electrochemical detection methods were developed for the detection of H_2O_2 based on peroxidase (or nanozyme) or using the strong oxidation of H_2O_2. Some representative approaches are summarized in Table 1. Hani et al. [32]. prepared Ce-MOF nanozyme, and constructed a fluorescence detection method for detecting H_2O_2 by the reaction of H_2O_2 with the Ce node of Ce-MOF. The detection range of H_2O_2 by this method was 200–1500 μM with an LOD of 10 μM. Wang et al. synthesized CeO QDs, and the introduction of QDs improved the charge transfer efficiency, thus enabling the electrochemical detection of H_2O_2. The method can detect and analyze H_2O_2 from 294 μM to 1.47 mM with an LOD of 26.5 μM [33]. These detection sensors achieve the quantitative detection of H_2O_2, but the process of probe preparation is complex, with harsh conditions and long preparation cycles. Zhang et al. established an H_2O_2 electrochemical detection platform based on Cat-HMFs/GCE, which achieved H_2O_2 detection from 100 μM to 3 mM with the LOD of 50 μM [34]. Biological enzymes have good catalytic activity, but the susceptibility of biological enzymes to deactivation limits their application. Thus, K.V. et al. cleverly designed an H_2O_2 detection paper chip by using the peroxidase property of chitosan. The linearity of the chitosan-based assay was found to be in the range of 10 μM to 10 mM with an LOD of 1.55 μM [35]. Compared with biological enzymes, chitosan has higher stability, easier storage and transportation, and a wider application range. In this work, a water-soluble detection probe was prepared by a relatively simple preparation method, and a more sensitive and rapid quantitative analysis of H_2O_2 was achieved without the introduction of other reagents. It has incomparable advantages in terms of detection sensitivity and ease of detection.

Table 1. Review of methods for detection of H_2O_2.

Materials	Synthetic Conditions	Methods	Wavelength	LOD	Detection Range	Linear/Logarithmic Relationship	Ref.
Ce-MOF	Solvothermal, 85 °C for 16 h	Fluorescence	λex at 356 nm, λem at 540 nm	10 μM	200–1500 μM	Linear	[32]
CeO-QDs	Calcined at 900 °C for 3 h, stirring for 24 h in water and reacted at 120 °C for a further 6 h.	GCE	——	26.5 μM	294 μM–1.47 mM	Linear	[33]
Cat-HMFs	ultrasonication at 25 °C for 30 min	GCE	——	50 μM	100 μM–3 mM	Linear	[34]
Chitosan	Waterthermal, 60 °C for 30 min	Colorimetric	λ_{Ab} = 652 nm	1.55 μM	10 μM–10 mM	Logarithmic	[35]
QDs	Waterthermal, refluxing for 2 h	Fluorescence	λex at 365 nm, λem at 610 nm	0.61 μM	1.23–300 μM	Logarithmic	This work

3.3. *Characterization of MnO_2 NSs*

The redox reaction between GSH and MnO_2 ($MnO_2 + 2GSH + 2H^+ \rightarrow Mn^{2+} + GSSG + 2H_2O$) is often used in biological applications. Inspired by this, MnO_2 is intended to react with GSH, the ligand of GSH-CdTe QDs, to quench its fluorescence. Among the various forms of MnO_2, MnO_2 NSs have been considered as probe elements due to their good adsorption properties. Before verifying the fluorescence quenching ability of MnO_2 NSs to GSH-CdTe QDs, the feasibility of MNO_2 nanosheets as probe elements was first evaluated (Figure 3). The few-layered MnO_2 nano-sheets with wrinkles and curling structures were observed by TEM image (Figure 3a,b), and it was found that the lattice fringe (d = 0.245 nm) corresponded to the (111) crystal plane of MnO_2 [36]. In addition, the optical properties of MnO_2 NSs were verified by UV-Vis absorption spectroscopy and fluorescence spectroscopy, respectively (Figure 3c,d), MnO_2 NSs has an obvious concentration-dependent characteristic absorption peak at 350 nm, which was attributed to the result of d-d electronic transitions of manganese ions in the [MnO_6] of layered MnO_2 NSs. As well, its absorbance value decreased linearly with the decrease in MnO_2 NSs concentration, which proved that the MnO_2 NSs were homogenous and stable. The above optical data supplemented the characterization of MnO_2 and also verified the feasibility of it as an optical signal component.

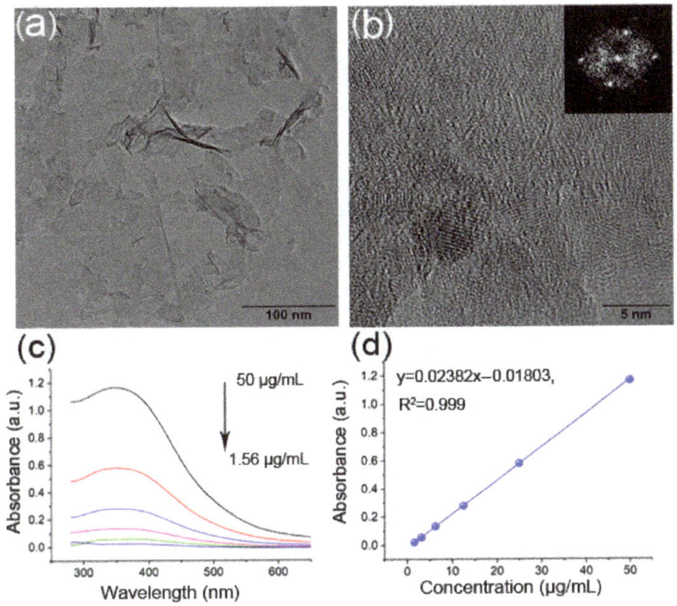

Figure 3. Characterization of MnO_2 NSs. TEM images (**a**) and HR-TEM image (Insert: Fourier diffractogram image) (**b**) of MnO_2 NSs; UV-Vis spectroscopy (**c**) and the linear relationship between Absorbance at 365 nm (**d**) versus 1.56–50 μg/mL of MnO_2 NSs.

3.4. Feasibility Analysis of MnO_2/GSH-CdTe QDs Fluorescent Sensing System

Fortunately, the phenomenon of fluorescence quenching of GSH-CdTe QDs by MnO_2 NSs was observed under UV light (Figure S2), and then the quenching performance was evaluated by fluorescence spectroscopy (Figure 4a). Similar to the quenching result of GSH-CdTe QDs by H_2O_2, the fluorescence value of GSH-CdTe QDs showed a logarithmic decrease trend rather than a linear trend with the increase of MnO_2 NSs. This result tentatively demonstrated that MnO_2 NSs and GSH-CdTe quantum dots can establish a fluorescence quenching system, and also proved that the system was not simply contributed by the fluorescence internal filtration effect (IFE). To verify this conclusion, two non-thiol ligand QDs were selected as fluorescent donor elements, and the results were shown in Figure 4b. The fluorescence of these two non-thiol ligands QDs both quenched as linear decrease trends with the increase of MnO_2 NSs, which was attributed to the IFE. In addition, the quenching efficiency of MnO_2 NSs for non-thiol capped QDs was much lower than that for GSH-CdTe QDs, when 10 μg/mL of MnO_2 NSs were added to GSH-CdTe QDs and non-thiol capped QDs with the same fluorescence intensity, the quenching rate of GSH-CdTe QDs (27.08%) was 2.2-fold higher than that of non-thiol capped QDs (8.4%). It was guessed that in addition to the IFE, the destruction of GSH ligands by MnO_2 NSs led to the disintegration of QDs, thereby increasing the quenching rate. The classic redox reaction between MnO_2 and GSH provided support for the construction of a highly sensitive MnO_2 NSs/GSH-CdTe QDs fluorescence quenching system.

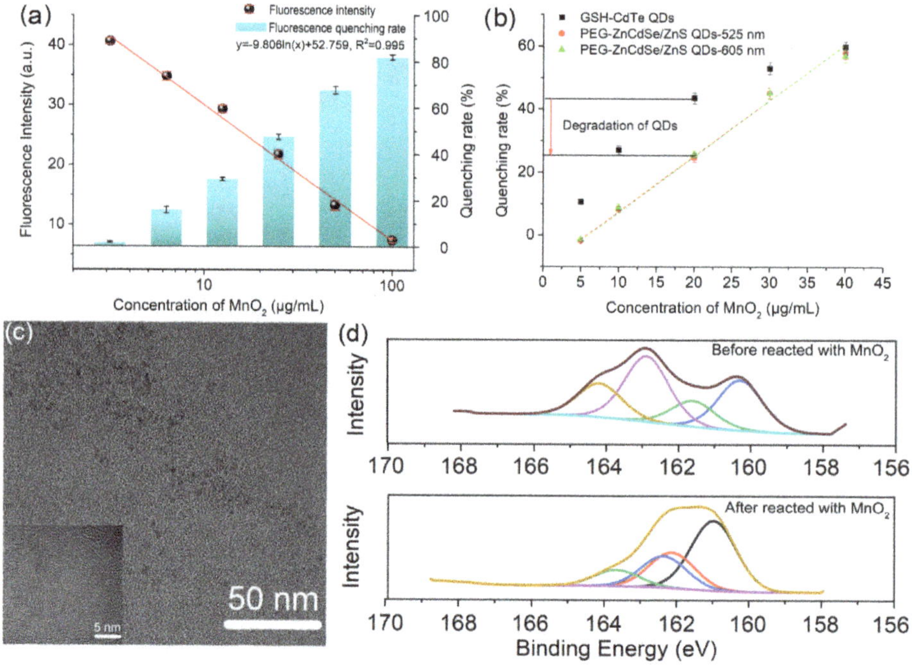

Figure 4. Verification of MnO_2 NSs/GSH-QDs quenching system. Fluorescence intensity and quenching rate of GSH-CdTe QDs after reaction with different concentrations of MnO_2 NSs (**a**); Comparison of the quenching ability of MnO_2 NSs to deferent QDs (**b**); TEM images and HR-TEM images (Insert) of MnO_2 NSs/GSH-QDs system (**c**); and XPS of GSH-CdTe QDs (S 2p) before and after reacted with MnO_2 NSs (**d**).

In order to further verify the above conclusions, MnO_2 NSs and GSH-CdTe QDs were mixed in water and their structures were characterized and evaluated by TEM and X-ray photoelectron spectroscopy (XPS). The TEM image of the mixture showed that the structures of MnO_2 NSs and QDs were significantly disintegrated, and the original morphology could no longer be observed (Figure 4c). At the same time, the results of H_2O_2 interacting with GSH-CdTe QDs showed that under the action of H_2O_2, the fluorescence of QDs decreased significantly (Figure 4a and Figure S4), and a logarithmic proportional relationship was obtained between the fluorescence intensity and the concentration of H_2O_2, which was consistent to the quenching trend of MnO_2/GSH-QDs quenching system. It further proved that the oxidation of sulfide (−2) was an important cause of fluorescence quenching of GSH-CdTe QDs [37]. The XPS results of GSH−CdTe QDs before and after reaction with MnO_2 NSs also showed clear changes in the binding energies of S2p. Before oxidation by MnO_2, the binding energies of S 2p (Figure 4d) at 161.2, 162.8 and 164.2 eV were ascribed to the sulfur in CdS, $CdTexS_{1-x}$ and sulfide (−2), and after oxidation by MnO_2, the binding energies of S 2p at 163.6 eV was appeared instead of 161.2, 162.8 and 164.2 eV, which ascribed to S_2^{2-} [31]. These results indicated that oxidation of the sulfide(−2) in the shell of CdS or $CdTexS_{1-x}$ by MnO_2 NSs, leads to the passivation layer being broken and effectively quenching the fluorescence [38]. In general, MnO_2 NSs and GSH-CdTe QDs have the ability to establish a fluorescence quenching system, and compared to most fluorescence quenching signals based solely on optical changes, this fluorescence quenching system had higher fluorescence quenching efficiency and specificity.

3.5. Establishment of Fluorescence Quenching Immunosensors (FQIS)

In addition to the above-mentioned optical properties, the surface adsorption properties of the sheet structure of MnO_2 NSs and the chemical covalent binding properties of the GSH-CdTe QDs provided the basis for the construction of immuno-probes (Scheme 1a,b). Before establishing the method, the specificity of the selected ENR-Ab was analyzed by indirect competition ELISA (Table S1). ENR-Ab did not recognize other quinolone antibiotics at all, except for a 1.8% cross-reactivity to ciprofloxacin, demonstrating the good specificity of ENR-Ab. As well, it is in agreement with previous reports from our group [39–41]. Through physical adsorption, the ENR detection antibody was coupled with MnO_2 NSs in a suitable pH environment to prepare the fluorescent receptor immuno-probe (MnO_2-Ab). OVA was coupled with the carboxyl/am group on the surface of the GSH-CdTe QDs by chemical bonding to prepare fluorescent donor probes (QDs-OVA). It was worth mentioning that the type of buffer during the preparation of QDs-OVA greatly affected the performance of the probe (Figure S5), QDs-OVA coupled in H_2O and PBS showed strong fluorescence, but the fluorescence intensity decreased significantly in HEPES and MES buffers, especially in MES buffers, the fluorescence of QDs was almost completely quenched (Figure S5A). In addition, the QDs-OVA conjugated in water and HEPES solution aggregated as a line on the NC membrane, while the conjugate prepared in ionic buffer (PBS) could not aggregate on the NC membrane, and only the conjugate prepared in H_2O was not washed out by the buffer during the chromatography. The above results demonstrate that the QDs-OVA coupling was successful in H_2O and could be successfully immobilized on the NC membrane. The fluorescence quenching immunosensors (FQISs) (Scheme 1c) were constructed based on the immunochromatography platform with QDs-OVA immobilizing on the surface of the NC membrane as the Control line (C line), and the mixture of QDs-OVA and ENR-antigen fixing under C line as Test line (T line). The details of the preparation and detection principle of FQISs were listed in Supplementary Information (SI).

During the working process of FQISs, the type of buffer was found to have significant impacts on the chromatography of MnO_2 NSs (Figure S6). The MnO_2-Ab probe aggregated in ionic buffers (PBS, PBST, $NaCO_3$-$NaHCO_3$ buffer) and could not be chromatographed on the NC membrane. However, it smoothly passed through the NC membrane and bound with the antigen at the T line in non-ionic buffers (MES, HEPES). It was guessed that the ionic solution destroyed the surface electronic environment of MnO_2 NSs, leading to the aggregation of MnO_2 NSs during the chromatography. The HEPES buffer (20 mM, Figure S7) was used for the establishment of FQISs, which was found to be more suitable for probe chromatography and antigen-antibody binding. Totals of 0, 0.05, 0.15, 0.45, and 1.35 ng/mL of ENR standards solution were mixed with 10 μL of MnO_2-Ab probe and added to the sample pad, after 5 min of chromatography, the detection results were observed (Figure 5). Under a UV lamp (Figure 4a), the MnO_2-Ab probe was enough for quenching the fluorescence of the T line when detecting ENR free solution. However, when detecting 0.05 ng/mL of ENR, some of the MnO_2-Ab probes were bound by ENR, and the remaining probes were not enough to quench all the fluorescence at the T line, so the fluorescence appeared. Thus, the LOD of these FQISs was 0.05 ng/mL of ENR in the HEPES buffer. However, the brown strip on the T line disappeared under the sunlight until the concentration of ERN reached 1.35 ng/mL (Figure 4b).

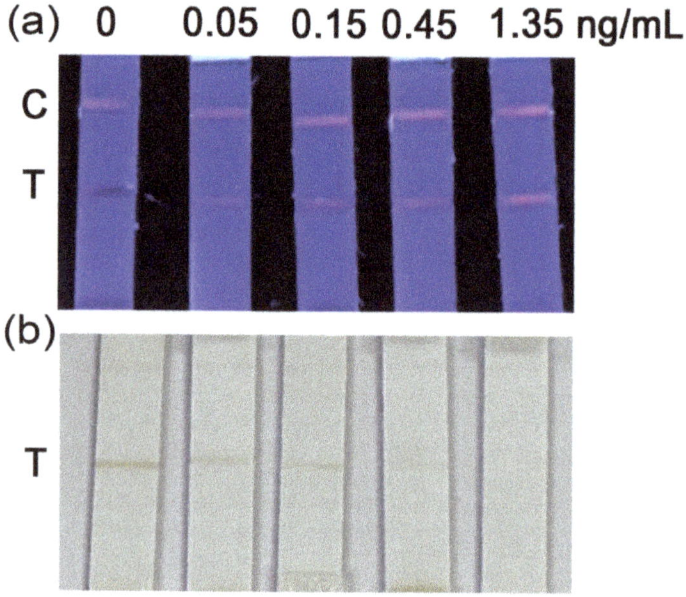

Figure 5. Detection of ENR using MnO_2 NSs-based fluorescent quenching immunosensors (FQISs). A series of concentrations of ENR (0, 0.05, 0.15, 0.45 and 1.35 ng/mL) was prepared in pH 7.4 HEPES (20 mM) and detected under (**a**) UV lamp and (**b**) natural light.

Tap water, milk and crucian carp from different regions or brands (n = 3) were used for detecting to verify the practicability of FQISs (The details of sample preparation were shown in Supplementary Information (SI), and each sample was detected three times), and they were verified as ENR free samples by liquid chromatography-mass spectrometry. Totals of 0, 0.05, 0.1 and 0.5 ng/mL of ENR were added into the tap water, 0, 0.1, 0.2 and 0.5 ng/mL of ENR were added into milk samples and 0, 0.25, 0.5, and 1 ng/g of ENR were added to the crucian carp samples respectively, after mixing and standing overnight at 4 °C, the targets were extracted and detected by FQISs, and the detection results were shown in Table 2. There was no false positive detection result when detecting tap water and milk, but only one weak fluorescent strip (weaker than the LOD of FQIS) appeared when detecting one of the crucian carp samples in three consecutive tests. The ELISA kits also did not detect the presence of ENR in this sample, even after adding 0.25 ng/g of ENR. Indicating that the ENR in this crucian carp sample was less than 0.25 ng/g and the above results proved that the proper sample preparation process greatly reduced the influence of the matrix and avoided false positive results. The fluorescence appeared clearly at the T line when detecting water samples with 0.05 ng/mL ENR addition, milk samples with 0.1 ng/mL ENR addition and crucian carp samples with 0.25 ng/g ENR addition, thus the LOD of ENR by FQISs was 0.05 ng/mL in water, 0.1 ng/mL in milk and 0.25 ng/g in crucian carp. In addition, the above detection results were also confirmed by commercial ELISA kits to verify the accuracy of FQISs. The results showed that the detection results of FQISs agree with ELISA kits, and FQISs achieved faster detection with 2 to 10-fold higher sensitivity for food samples detection, respectively (Table S2).

Table 2. Comparison of the developed FISs and the commercial ELISA kit for the detection of ENR in tap water and crucian carp samples.

Sample	Conc. (ng/g(mL))	Detection Results (n = 3)	Observed Results	ELISA Kits (n = 3) (ng/g(mL))
Tap water	0	−; −; − [a]	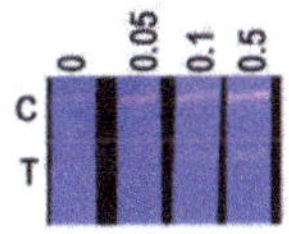	ND [b]
	0.05	+; + [c]; ± [d]		ND
	0.1	+; +; +		ND
	0.5	+; +; +		0.47 ± 0.04
Milk	0	−; −; −	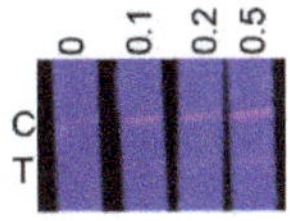	ND
	0.1	+; +; ±		ND
	0.2	+; +; +		ND
	0.5	+; +; +		0.42 ± 0.07
Crucian carp	0	−; −; ±	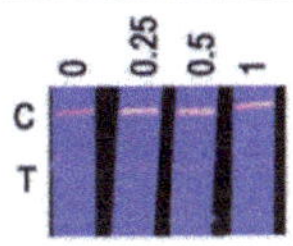	ND
	0.25	+; +; +		ND
	0.5	+; +; +		0.39 ± 0.08
	1	+; +; +		0.86 ± 0.11

[a] ND: indicates no ENR detection or negative test result. [b] −: indicates a negative test result. [c]+: indicates a positive test result [d] ±: indicates a weakly positive test result.

Benefiting from the specific recognition performance of antigen-antibody, immunosensors are considered to be one kind of the most effective rapid detection and analysis tools. However, limited by the affinity of the antibody itself, the lower detection sensitivity of traditional immunoassays has not been improved. More and more functional materials are introduced into immunoassays as signal elements to improve sensitivity and other detection performance. Among them, the fluorescence signal has higher signal resolution efficiency than the colorimetric signal, and further, the fluorescence quenching signal performs better. In addition, differences in antibody loading rates and signal generation mechanisms of signal elements also have significant impacts on detection sensitivity. Reviewing the different immunochromatographic sensing systems established with the same detection antibody by our group, the above conclusions have been verified (Table 3).

Table 3. An overview of immunosensors with the same antibody for determination of ENR.

Signal Type	Signal Element	Detection Limit (ng/mL)	Ref.
Colorimetric	AuNPs	5	[39]
	Black phosphorus-Au nanocomposite	0.5	[41]
Fluorescence	Dyed polymer microsphere	1	[39]
	PEG-QDs	1	[39]
Fluorescence quenching	AuNPs/PEG-QDs	0.25	[40]
	AgNPs/carbon dots(CDs)	0.1	[40]
Chemical degradation + fluorescence quenching	MnO_2 NSs/GSH-QDs	0.05	This work

The fluorescence system has made a contribution to improving sensitivity due to its stronger ability to resist background and matrix interference, thus the detection limit is decreased by 5-fold than the AuNPs system [40]. However, limited by lower antibody coupling efficiency and the "turn off" signal output mode, the sensitivity of fluorescence analysis needs to be further improved. Compared with the above two signal systems, the fluorescence quenching systems with "turn on" signal output mode for detecting small molecule targets take the appearance of fluorescence signal as the criterion of detection limit, so it has the highest detection sensitivity [40]. In the process of constructing fluorescence quenching systems, the matching of fluorescence donor and acceptor probes is the most important factor. Compared with AuNPs/PEG-QDs fluorescence quenching system, the higher fluorescence quenching efficiency of Ag nanoparticles (AgNPs) to CDs determines that the AgNPs/CDs quenching system has the ability to increase detection sensitivity. However, the blue fluorescence of CDs is similar to that of most biological samples or background light, so that the application of AgNPs/CDs quenching system in the detection of complex samples is limited.

It is worth mentioning that, compared with 0D nanomaterials, 2D nanomaterials show unique advantages in immunosensing analysis. For example, in the colorimetric system, 2D black phosphorus nanosheets (BPNSs) have better optical absorption and adsorption capacity than 0D gold nanoparticles (AuNPs). Therefore, the detection limit of immunosensors based on BP-Au nanocomposites is 10-fold lower than that of AuNPs-based immunosensors, and even lower than that of fluorescence detection systems [41]. In view of this, the organic combination of two-dimensional nanomaterials and fluorescence quenching signal system is bound to further improve the signal sensitivity. Two-dimensional MnO_2 NSs were cleverly selected as fluorescent acceptor probes, and in addition to their excellent optical properties and adsorption properties, their chemical stability was stronger than that of BPNSs. More surprisingly, MnO_2 NSs possessed unique and incomparable redox properties. Based on this, water-soluble GSH-CdTe QDs were selected as fluorescent donor probes, and a fluorescence quenching system of MnO_2 NSs/GSH-CdTe QDs was constructed. In addition to fluorescence internal filtration, the irreversible disintegration of QDs due to the redox reaction between MnO_2 and GSH further enhances the fluorescence quenching efficiency and specificity of this quenching system. Reassuringly, the novel fluorescence quenching system based on 2D MnO_2 NSs brings amazing sensitivity for immunosensing assays. Compared with the AuNPs-based colorimetric detection results, the FQISs are 100-fold more sensitive when used for ENR detection, and approached the detection limit of photothermal quantitative detection. The construction of The FQISs lays the foundation for the in-depth study of the new fluorescence quenching system and its application in immunosensing, and is expected to promote the application of immunosensing analysis in the highly sensitive detection of trace pollutants. However, the work at this stage is still in the detection of antibiotic residues in primarily processed foods, and the current pre-treatment method is not yet able to achieve the extraction of antibiotics in deeply processed foods, and further exploration of the pre-treatment method for deeply processed food samples is the direction of subsequent research. In addition, the follow-up work will also actively advance this FQIS to a quantitative detection sensor to make it more accurate.

4. Conclusions

Water-soluble GSH-CdTe QDs with an average particle size of 3.2 nm and a maximum emission wavelength of 615 nm were successfully prepared by a solvothermal method. their H_2O_2 responsiveness was surprisingly found, and a highly sensitive and rapid label-free H_2O_2 fluorescence detection method was constructed based on it. When used for the detection of H_2O_2 in milk, the red fluorescence of GSH-CdTe QDs showed the advantage of being unaffected by the fluorescence of the milk matrix, which increased the detection accuracy and improved the detection sensitivity at the same time. The LOD was 0.61 μM of H_2O_2 in water and 68 μM of H_2O_2 in milk. In addition, the MnO_2 NSs/GSH-CdTe QDs fluorescence quenching system was constructed. At 10 μg/mL of MnO_2 NSs, the quenching rate of GSH-CdTe QDs was 2.2-fold higher than that of other ligand-capped QDs due to the oxidation of sulfide (−2) in the CdS or $CdTexS_{1-x}$ shells by MnO_2 NSs. Based on the MnO_2 NSs/GSH-CdTe QDs fluorescence quenching system, we developed a highly sensitive fluorescence quenching immunosensor (FQIS). Targeting enrofloxacin, the FQISs achieved high sensitivity detection with a detection limit of 0.05 ng/mL in HEPES buffer, without the need for expensive detection equipment. For the detection of environmental water samples and food samples of animal origin, the results were consistent with commercially available ELISA kits. Benefiting from the MnO_2 NSs/GSH-CdTe QDs fluorescence quenching system, the FQISs presented in this work have higher sensitivity than other quenching systems.

Using water-soluble GSH-CdTe QDs as fluorescent probes, rapid and highly sensitive detection and analysis of contaminant residues such as additives and antibiotics in the environment and food was achieved. These promising strategies are expected to be used for the detection of other high-sensitivity biomolecules.

Supplementary Materials: The following supporting information can be downloaded at: https://www.mdpi.com/article/10.3390/foods12010062/s1, Methods: 1. Preparation of enrofloxacin coating antigen (ENR-OVA); 2. Preparation of FQISs; 3. Detection principle; 4. Sample preparation. Figures: Figure S1: Particle size distribution of GSH-CdTe QDs; Figure S2: Figure S2 Visualized results of GSH-CdTe QDs with or without MnO2 NSs under (A) Sun light and (B) UV light; Figure S3 The selectivity of GSH-CdTe QDs; Figure S4: Plots of fluorescence values versus H_2O_2 concentrations; Figure S5: Optimization of coupling conditions for fluorescent sensing probes; Figure S6: Chromatography results of fluorescent quenching probes in different buffers. From left to right: 20 mM of PBS, 30 mM of MES, $NaCO_3$-$NaHCO_3$ buffer, PBST, 20 mM of HEPES; Figure S7: Chromatography results of fluorescent quenching probes in different concentration of HEPES. From left to right: 10 mM, 20 mM, 50 mM and 100 mM of HEPES. Tables: Table S1 Specificity analysis of ENR-Ab, Table S2 Comparison of the analytical performance of the FQISs with the commercial ELISA test kit to detect enrofloxacin in animal origin food samples.

Author Contributions: S.L.: Conceptualization, Methodology, Funding acquisition, Writing—original draft. L.N.: Methodology, Data curation, Writing—original draft. L.H.: Data curation. W.W.: Data curation. J.W.: Methodology. S.W.: Conceptualization, Methodology, Supervision, Writing—review and editing. All authors have read and agreed to the published version of the manuscript.

Funding: This work was supported by the National Natural Science Foundation of China (No. 32102068) and the 68th China Postdoctoral Science Foundation (Grant No.2020M680870).

Institutional Review Board Statement: Not applicable.

Informed Consent Statement: Not applicable.

Data Availability Statement: The data presented in this study are available on request from the corresponding author.

Conflicts of Interest: The authors declare no conflict of interest.

References

1. Kemper, N. Veterinary antibiotics in the aquatic and terrestrial environment. *Ecol. Indic.* **2008**, *8*, 1–13. [CrossRef]
2. King, T.; Cole, M.; Farber, J.M.; Eisenbrand, G.; Zabaras, D.; Fox, E.M.; Hill, J.P. Food safety for food security: Relationship between global megatrends and developments in food safety. *Trends Food Sci. Technol.* **2017**, *68*, 160–175. [CrossRef]
3. Lv, M.; Liu, Y.; Geng, J.; Kou, X.; Xin, Z.; Yang, D. Engineering nanomaterials-based biosensors for food safety detection. *Biosens. Bioelectron.* **2018**, *106*, 122–128. [CrossRef] [PubMed]
4. Shao, B.; Li, H.; Shen, J.; Wu, Y. Nontargeted Detection Methods for Food Safety and Integrity. *Annu. Rev. Food Sci. Technol.* **2019**, *10*, 429–455. [CrossRef]
5. Zhang, G.; Deng, S.; Fang, B.; Zhang, G.; Lai, X.; Su, L.; He, W.; Lai, W. Lateral flow immunoassay based on polydopa-mine-coated metal–organic framework for the visual detection of enrofloxacin in milk. *Anal. Bioanal. Chem.* **2022**, *414*, 7315–7323. [CrossRef]
6. Zhang, Z.; Liu, Q.; Zhang, M.; You, F.; Hao, N.; Ding, C.; Wang, K. Simultaneous detection of enrofloxacin and ciprofloxacin in milk using a bias potentials controlling-based photoelectrochemical aptasensor. *J. Hazard. Mater.* **2021**, *416*, 125988. [CrossRef]
7. Aymard, C.; Kanso, H.; Serrano, M.J.; Pagán, R.; Noguer, T.; Istamboulie, G. Development of a new dual electrochemical immunosensor for a rapid and sensitive detection of enrofloxacin in meat samples. *Food Chem.* **2022**, *370*, 131016. [CrossRef]
8. Gill, R.; Bahshi, L.; Freeman, R.; Willner, I. Optical Detection of Glucose and Acetylcholine Esterase Inhibitors by H2O2-Sensitive CdSe/ZnS Quantum Dots. *Angew. Chem.* **2008**, *120*, 1700–1703. [CrossRef]
9. Lima, L.S.; Rossini, E.L.; Pezza, L.; Pezza, H.R. Bioactive paper platform for detection of hydrogen peroxide in milk. *Spectrochim. Acta Part A Mol. Biomol. Spectrosc.* **2020**, *227*, 117774. [CrossRef]
10. Ren, H.-X.; Qian, Z.-J.; Li, M.; Peng, C.-F.; Wang, Z.-P.; Wei, X.-L.; Xu, J.-G. Mesoporous silica-loaded gold nanocluster with enhanced fluorescence and ratiometric fluorescent detection of thiram in foods. *Mikrochim. Acta* **2021**, *188*, 363. [CrossRef]
11. Zhang, Y.; Liu, R.; Hassan, M.; Li, H.; Ouyang, Q.; Chen, Q. Fluorescence resonance energy transfer-based aptasensor for sensitive detection of kanamycin in food. *Spectrochim. Acta Part A Mol. Biomol. Spectrosc.* **2021**, *262*, 120147. [CrossRef] [PubMed]
12. Tseng, W.-B.; Rau, J.-Y.; Chiou, H.-C.; Tseng, W.-L. Synthesis of gold nanoclusters-loaded lysozyme nanoparticles for rati-ometric fluorescent detection of cyanide in tap water, cyanogenic glycoside-containing plants, and soils. *Environ. Res.* **2022**, *207*, 112144. [CrossRef] [PubMed]

13. Zhang, W.; Zhong, H.; Zhao, P.; Shen, A.; Li, H.; Liu, X. Carbon quantum dot fluorescent probes for food safety detection: Progress, opportunities and challenges. *Food Control.* **2022**, *133*, 108591. [CrossRef]
14. Wagner, A.M.; Knipe, J.M.; Orive, G.; Peppas, N.A. Quantum dots in biomedical applications. *Acta Biomater.* **2019**, *94*, 44–63. [CrossRef] [PubMed]
15. Liu, X.; Sun, Y.; Lin, X.; Pan, X.; Wu, Z.; Gai, H. Digital Duplex Homogeneous Immunoassay by Counting Immunocomplex Labeled with Quantum Dots. *Anal. Chem.* **2021**, *93*, 3089–3095. [CrossRef] [PubMed]
16. Jones, L.O.; Mosquera, M.A.; Jiang, Y.; Weiss, E.A.; Schatz, G.C.; Ratner, M.A. Thermodynamics and Mechanism of a Photocatalyzed Stereoselective [2 + 2] Cycloaddition on a CdSe Quantum Dot. *J. Am. Chem. Soc.* **2020**, *142*, 15488–15495. [CrossRef]
17. Zhou, W.; Coleman, J.J. Semiconductor quantum dots. *Curr. Opin. Solid State Mater. Sci.* **2016**, *20*, 352–360. [CrossRef]
18. Jiao, X.; Zhou, Y.; Zhao, D.; Pang, D.; Wang, C.; Du, H.; Wen, Y.; Zhang, X. An indirect ELISA-inspired dual-channel fluorescent immunoassay based on MPA-capped CdTe/ZnS QDs. *Anal. Bioanal. Chem.* **2019**, *411*, 5437–5444. [CrossRef]
19. Zhu, Y.; Chao, J.; Zhu, F.; Zhu, N.; Zhang, Q.; Gyimah, E.; Yakubu, S.; Zou, Y.; Zhang, Z. Ratiometric fluorescence immunoassay based on FAM-DNA–functionalized CdSe/ZnS QDs for the sensitive detection of tetrabromobisphenol A in foodstuff and the environment. *Anal. Bioanal. Chem.* **2020**, *412*, 3605–3613. [CrossRef]
20. Dou, X.; Dai, H.; Twardowska, I.; Wei, S. Hyperaccumulation of Cd by Rorippa globosa (Turcz.) Thell. from soil enriched with different Cd compounds, and impact of soil amendment with glutathione (GSH) on the hyperaccumulation efficiency. *Environ. Pollut.* **2019**, *255*, 113270. [CrossRef]
21. Qian, H.; Dong, C.; Weng, J.; Ren, J. Facile One-Pot Synthesis of Luminescent, Water-Soluble, and Biocompatible Glutathione-Coated CdTe Nanocrystals. *Small* **2006**, *2*, 747–751. [CrossRef] [PubMed]
22. Liu, Y.-F.; Yu, J.-S. Selective synthesis of CdTe and high luminescence CdTe/CdS quantum dots: The effect of ligands. *J. Colloid Interface Sci.* **2009**, *333*, 690–698. [CrossRef] [PubMed]
23. Han, L.; Liu, S.G.; Liang, J.Y.; Li, N.B.; Luo, H.Q. Free-label dual-signal responsive optical sensor by combining resonance Rayleigh scattering and colorimetry for sensitive detection of glutathione based on ultrathin MnO2 nanoflakes. *Sensors Actuators B Chem.* **2019**, *288*, 195–201. [CrossRef]
24. Zhang, J.; Yang, H.; Pan, S.; Liu, H.; Hu, X. A novel "off-on-off" fluorescent-nanoprobe based on B, N co-doped carbon dots and MnO2 nanosheets for sensitive detection of GSH and Ag+. Spectrochim. *Acta Part A Mol. Biomol. Spectrosc.* **2020**, *244*, 118831. [CrossRef]
25. Fan, D.; Shang, C.; Gu, W.; Wang, E.; Dong, S. Introducing Ratiometric Fluorescence to MnO2 Nanosheet-Based Biosensing: A Simple, Label-Free Ratiometric Fluorescent Sensor Programmed by Cascade Logic Circuit for Ultrasensitive GSH Detection. *ACS Appl. Mater. Interfaces* **2017**, *9*, 25870–25877. [CrossRef]
26. Liu, J.; Meng, L.; Fei, Z.; Dyson, P.J.; Jing, X.; Liu, X. MnO 2 nanosheets as an artificial enzyme to mimic oxidase for rapid and sensitive detection of glutathione. *Biosens. Bioelectron.* **2017**, *90*, 69–74. [CrossRef]
27. Zhang, S.; Sunami, Y.; Hashimoto, H. Mini Review: Nanosheet Technology towards Biomedical Application. *Nanomaterials* **2017**, *7*, 246. [CrossRef]
28. Chen, J.; Meng, H.; Tian, Y.; Yang, R.; Du, D.; Li, Z.; Qu, L.; Lin, Y. Recent advances in functionalized MnO2 nanosheets for biosensing and biomedicine applications. *Nanoscale Horizons* **2018**, *4*, 321–338. [CrossRef]
29. Yan, Y.; Sun, J.; Zhang, K.; Zhu, H.; Yu, H.; Sun, M.; Huang, D.; Wang, S. Visualizing Gaseous Nitrogen Dioxide by Ratiometric Fluorescence of Carbon Nanodots–Quantum Dots Hybrid. *Anal. Chem.* **2015**, *87*, 2087–2093. [CrossRef]
30. Gendloff, E.; Casale, W.; Ram, B.; Tai, J.; Pestka, J.; Hart, L. Hapten-protein conjugates prepared by the mixed anhydride method: Cross-reactive antibodies in heterologous antisera. *J. Immunol. Methods* **1986**, *92*, 15–20. [CrossRef]
31. Li, H.; Lu, W.; Song, B.; Zhou, J.; Zhao, G.; Han, G. The design of Mn2+&Co2+ co-doped CdTe quantum dot sensitized solar cells with much higher efficiency. *RSC Adv.* **2020**, *10*, 35701–35708. [CrossRef] [PubMed]
32. Abdelhamid, H.N.; Sharmoukh, W. Intrinsic catalase-mimicking MOFzyme for sensitive detection of hydrogen peroxide and ferricions. *Microchem. J.* **2021**, *163*, 105873. [CrossRef]
33. Wang, Y.; Qian, J.; Chen, Z.; Wang, C.; Liu, C.; Yang, Y.; Wu, Z.; Song, Y. CeO2 quantum dots modified electrode for detecting hydrogen peroxide. *Inorg. Chem. Commun.* **2019**, *101*, 62–68. [CrossRef]
34. Zhang, M.; Chen, M.; Liu, Y.; Wang, Y.; Tang, J. Catalase-inorganic hybrid microflowers modified glassy carbon electrode for amperometric detection of hydrogen peroxide. *Mater. Lett.* **2019**, *243*, 9–12. [CrossRef]
35. Ragavan, K.; Ahmed, S.R.; Weng, X.; Neethirajan, S. Chitosan as a peroxidase mimic: Paper based sensor for the detection of hydrogen peroxide. *Sensors Actuators B Chem.* **2018**, *272*, 8–13. [CrossRef]
36. Jia, H.; Cai, Y.; Lin, J.; Liang, H.; Qi, J.; Cao, J.; Feng, J.; Fei, W. Heterostructural Graphene Quantum Dot/MnO2 Nanosheets toward High-Potential Window Electrodes for High-Performance Supercapacitors. *Adv. Sci.* **2018**, *5*, 1700887. [CrossRef]
37. Ngamchuea, K.; Batchelor-McAuley, C.; Compton, R.G. The Copper(II)-Catalyzed Oxidation of Glutathione. *Chem. A Eur. J.* **2016**, *22*, 15937–15944. [CrossRef]
38. Aldana, J.; Wang, Y.A.; Peng, X. Photochemical Instability of CdSe Nanocrystals Coated by Hydrophilic Thiols. *J. Am. Chem. Soc.* **2001**, *123*, 8844–8850. [CrossRef]
39. Sheng, W.; Li, S.; Liu, Y.; Wang, J.; Zhang, Y.; Wang, S. Visual and rapid lateral flow immunochromatographic assay for enrofloxacin using dyed polymer microspheres and quantum dots. *Microchim. Acta* **2017**, *184*, 4313–4321. [CrossRef]

40. Li, S.; Wang, Y.; Mu, X.; Sheng, W.; Wang, J.; Wang, S. Two fluorescence quenching immunochromatographic assays based on carbon dots and quantum dots as donor probes for the determination of enrofloxacin. *Anal. Methods* **2019**, *11*, 2378–2384. [CrossRef]
41. Li, S.; Zhang, Y.; Wen, W.; Sheng, W.; Wang, J.; Wang, S.; Wang, J. A high-sensitivity thermal analysis immunochromatographic sensor based on au nanoparticle-enhanced two-dimensional black phosphorus photothermal-sensing materials. *Biosens. Bioelectron.* **2019**, *133*, 223–229. [CrossRef] [PubMed]

foods

MDPI

Article

Preparation of Magnetic Metal-Organic Frameworks@Molecularly Imprinted Nanoparticles for Specific Extraction and Enrichment of Bisphenol A in Food

Qi Zhang [1], Haiyang Wang [1], Yongju Zhang [2], Zhixiang Xu [1] and Longhua Xu [1,*]

1 Key Laboratory of Food Processing Technology and Quality Control in Shandong Province, College of Food Science and Engineering, Shandong Agricultural University, Tai'an 271018, China; 15650452276@163.com (Q.Z.); ocean5416@163.com (H.W.); zhixiangxu@sina.com (Z.X.)
2 Shandong Institute for Product Quality Inspection, Jinan 250102, China; zhangyongju144@163.com
* Correspondence: longhuaxu@sdau.edu.cn; Tel.: +86-538-8246029

Abstract: Metal-organic frameworks (MOFs) with systematically tailored structures have been suggested as promising precursors to the preparation of diverse functional materials. Herein, a facile and versatile layer-by-layer strategy without any special surface modifications has been proposed for the preparation of magnetic metal-organic frameworks (MMOFs) supported molecularly imprinted polymer nanoparticles (MMOFs@MIP), which are based on a magnetically susceptible core conjugated with an imidazole-derived self-assembled layer and a silane-based imprinted shell. The obtained MMOFs@MIPs, which integrated the advantages of Fe_3O_4, MOFs, and MIPs, were characterized and exhibited good magnetic properties, a rapid mass transfer rate, and an excellent adsorption selectivity as well as capacity for the targeted molecular - bisphenol A (BPA). Moreover, the MMOFs@MIPs were employed as adsorbents in magnetic solid phase extraction (MSPE) to selectively bind and rapidly separate BPA from real samples with satisfactory recoveries ranging from 88.3% to 92.3%. More importantly, the desirable reusability of MMOFs@MIP was also evaluated, and the recoveries still maintained above 88.0% even after five re-use cycles. Furthermore, combined with high-performance liquid chromatography (HPLC) analysis, a novel MSPE-HPLC method was developed, enabling the highly selective and sensitive detection of BPA in a wide linear range of 0.5–5000 $\mu g\ L^{-1}$ with a low limit of detection (LOD) of 0.1 $\mu g\ L^{-1}$. This work contributes a promising method for constructing various functional nanoparticles @MOFs@MIP hybrid materials for applications in many different fields.

Keywords: magnetic solid phase extraction; metal-organic framework; molecularly imprinted polymer; bisphenol A

Citation: Zhang, Q.; Wang, H.; Zhang, Y.; Xu, Z.; Xu, L. Preparation of Magnetic Metal-Organic Frameworks@Molecularly Imprinted Nanoparticles for Specific Extraction and Enrichment of Bisphenol A in Food. *Foods* **2022**, *11*, 1408. https://doi.org/10.3390/foods11101408

Academic Editors: Federico Casanova and Beatriz Gullon

Received: 9 March 2022
Accepted: 7 May 2022
Published: 12 May 2022

1. Introduction

Bisphenol A is an important industrial monomer that is commonly used as a surfactant and a plasticizer in the manufacturing of polycarbonate plastics and epoxy resins and is applied to the coating of metal surfaces in contact with food and thermal paper [1]. The incomplete polymerization or polymer degradation of BPA allow it to easily migrate and be widely distributed in environmental matrices and food samples. BPA has been proved to be a representative endocrine disruptor that can cause serious damage to the reproductive, nervous, and immune systems and is closely related to many malignant tumors even with a low exposure dose [2,3]. Due to its health risks for humans and other organisms, BPA pollution has received tremendous attention worldwide. While limited to a low concentration, developing low-cost and efficient adsorption materials for monitoring BPA in the environment and food samples is of great significance.

Metal-organic frameworks (MOFs) with systematically tailored structures have been suggested as promising precursors to the preparation of diverse functional materials.

Magnetic metal-organic frameworks (MMOFs), which inherit the advantages of magnetic materials and MOFs, have attracted enormous attention in analytical chemistry due to their easy access, super-paramagnetism, rapid adsorption/separation, and reusability. To date, numerous MMOFs have been synthesized using various approaches including embedding, encapsulation, mixing, or layer-by-layer assembly [4], and their intriguing properties facilitated their application as effective adsorbents in sample collection and pre-enrichment, solid-phase extraction, and solid-phase microextraction in recent years. For instance, Yan and co-workers synthesized magnetic MIL-101 microcrystals by physically mixing MIL-101 and silica-coated Fe_3O_4 microparticles under ultrasonication and then used the resulting particles for magnetic solid-phase extraction of trace polycyclic aromatic hydrocarbons in water samples [5]. Subsequently, Li's team reported an embedding method for the fabrication of hybrid MMOF-5s via chemical covalent bonding between amino functionalized Fe_3O_4 nanoparticles and the surface of an MOF-5 for magnetic separation and the enrichment of polycyclic aromatic hydrocarbons and gibberellic acid from environmental, food, and plant samples [6]. An encapsulation strategy was adopted for the synthesis of a core–shell structured MMOF nanocomposite (Fe_3O_4@ZIF-8) for the elimination of U(VI) and Eu(III) from the environment [7]. The particles were synthesized by nucleation through PSS-modified Fe_3O_4 with a negative charge to attract Zn^{2+} cations to from a ZIF-8 layer. Fe_3O_4@AMCA-MIL53(Al) for the removal of U(VI) and Th(IV) metal ions from aqueous environments was also obtained through this strategy by using alkaline co-precipitation of $FeCl_2$ and $FeCl_3$ in the presence of AMCA-MIL53(Al) [8]. Recently, the layer-by-layer assembly approach has received considerable attention and has been widely employed to obtain well-defined core–shell-structured MMOFs. For example, Chen et al. synthesized Fe_3O_4@MIL-100 (Fe) for the elective capture of phosphopeptides by adding a MOF shell onto Fe_3O_4 nanoparticles [9]. Zhang et al. prepared a magnetic bimetallic metal-organic framework with Zr−O and Ti−O clusters for global phosphopeptide enrichment via the coordination of metal ions with carboxyl groups [10].

Despite the fact that a considerable quantity of MMOFs have been developed and exhibit unparalleled advantages as sorbents for the concentration and separation of trace analytes, their intractable or insufficient selectivity is still a great challenge. In order to solve this problem, numerous attempts have been made to chemical post-modification of the as-prepared MMOFs. For example, Ke et al. designed thiol-functionalized Fe_3O_4@MOF for the selective removal of Hg^{2+} and Pb^{2+} from wastewater [11], while Xu et al. employed a beta-cyclodextrin-functionalized MMOF to selectively extract prochloraz and triazole fungicides from vegetable samples [12].

However, the selectivity for a specific target that only depends on functional group modification is not highly developed, especially in a complex sample matrix. Thus, developing a simple and generally applicable methodology to improve the selectivity of MMOFs is of great importance. Fortunately, this possibility was enabled by molecular imprinting technology, which has been proven as an efficient and straightforward approach to producing artificial antibody-like materials with specific molecular-recognition sites [13–17]. The resulting molecularly imprinted polymers have been employed in MOF-based adsorption materials to improve their selectivity [18–20].

Inspired by these, we demonstrated a facile and versatile strategy for the synthesis of magnetic metal-organic frameworks with molecularly imprinted nanoparticles via an efficient sol–gel molecular imprinting process in the presence of MMOFs. The fabricated magnetic metal-organic frameworks @molecularly imprinted nanoparticles (MMOFs@MIP and Fe_3O_4@ZIF-8@MIP) presented a well-bedded core–shell structure, excellent magnetism, good reusability, fast adsorption, and high selectivity for the target molecule. It was successfully used as an adsorbent to extract and concentrate BPA, which is an endocrine disrupter that can be frequently found in environmental media or packaged foods. Furthermore, combined with HPLC analysis, the Fe_3O_4@ZIF-8@MIPs were employed as an absorbent in magnetic solid phase extraction (MSPE) for the highly selective and sensitive detection of BPA in food samples. The developed MSPE-HPLC method has a wide linear range of

0.5–5000 μg L^{-1} and a low LOD of 0.1 μg L^{-1} (S/N = 3). At three concentration levels of 0.5, 1.0, and 10 μg L^{-1}, the satisfactory recoveries ranging from 88.3% to 92.3% were obtained in uncontaminated lemon juice, canned hawthorn, and mineral water samples, indicating the excellent ability of the prepared Fe_3O_4@ZIF-8@MIP to recognize BPA in food samples and its potential for application in BPA detection.

2. Materials and Methods

2.1. Chemicals

3-aminopropyltriethoxysilane (APTES, 98%) and tetraethoxysilane (TEOS, 99%) were purchased from Shanghai Macklin Biochemical Co., Ltd. (Shanghai, China). Bisphenol A (BPA, 99%), bisphenol B (BPB, 99.5%), bisphenol F (BPF, 99%), bisphenol S (BPS, 99%), bisphenol AF (BPAF, 98%), resorcinol (HQ, 99%), and phenol (P, 99.5%) were obtained from Aladdin Industrial Corporation (Shanghai, China). Unless noted otherwise, all chemicals were used as received.

2.2. Preparation of Core–Shell MMOF@MIP (Fe_3O_4@ZIF-8@MIP)

An Fe_3O_4@ZIF-8 was first synthesized using a gentle one-pot self-assembly strategy from a previous report [21,22]. An Fe_3O_4@ZIF-8@MIP was fabricated as illustrated in Scheme 1. The pre-assembly solution was prepared by mixing the template BPA (1 mmol) and the functional monomer APTES (4 mmol) in ethanol (30 mL) under gentle stirring for 1 h. Next, Fe_3O_4@ZIF-8 (150 mg) as a support was dispersed into the preassembly solution and stirred for 0.5 h. Subsequently, the cross-linker TEOS (12 mmol) and catalyzer HCl (1 mL, 1 mol L^{-1}) were successively added dropwise. The pre-polymerization mixture was placed at room temperature for 1 h and then incubated in a water bath at 60 °C for 10 h. The resulting material was washed via Soxhlet extraction in acetic acid and methanol (1:9, v/v) until no template BPA was detectable by HPLC. Finally, the product was washed with methanol and dried under a vacuum to obtain Fe_3O_4@ZIF-8@MIP.

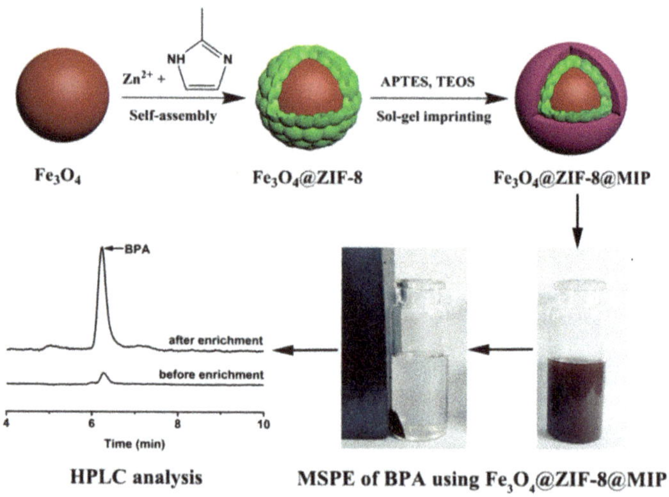

Scheme 1. Schematic illustration of the fabrication and application of Fe_3O_4@ZIF-8@MIP for extraction and enrichment of bisphenol A.

As a control, a non-imprinted Fe_3O_4@ZIF-8 (Fe_3O_4@ZIF-8@NIP) and a molecularly imprinted polymer coated magnetic Fe_3O_4 (Fe_3O_4@MIP) were also prepared following the same procedure but in the absence of the template BPA and using Fe_3O_4 as a support instead of Fe_3O_4@ZIF-8.

2.3. *Extraction and Detection of BPA in Real Samples*

2.3.1. Sample Pretreatment

The lemon juice, canned hawthorn, and mineral water samples were purchased from a local supermarket (Tai'an, China). A reasonable quantity of the aqueous phase in contact with canned foods was filtered through 0.45 μm glass fiber membrane syringe filters; the pH was adjusted to 7 by the dropwise addition of NaOH (1 M) and then stored at 4 °C for further use.

2.3.2. MSPE of BPA from Samples Using MMOF@MIP

The as-synthesized Fe_3O_4@ZIF-8@MIP was used as an absorbent for the magnetic solid phase extraction of BPA. Briefly, Fe_3O_4@ZIF-8@MIP P (0.5 mg) was added into a volumetric flask and rinsed in sequence with methanol and water. Then, the MMIPMs were separated under an external magnetic field and the supernatant was discarded. Subsequently, 50 mL of the sample extract solution was added into the volumetric flask and mixed via mechanically shaking for 30 min at room temperature, followed by separation with a magnet, and then eluted with a mixture of methanol and acetic acid (3 mL, v/v = 9:1) under ultrasonic treatment for 30 s. The collected eluent was evaporated to near dryness at a reduced pressure at 50 °C and re-dissolved with methanol to 1 mL for HPLC analysis.

2.3.3. HPLC Analysis

The detection of BPA was conducted via a Shimadzu LC-20AT HPLC system, which consisted of an LC-20AT pump, an RF-10AXL fluorescence detector set to an excitation wavelength of 227 nm and emission wavelength of 313 nm, an SIL-20A automatic sampler with a 20 μL injection loop, an LC workstation for data collection, a CTO-20A column oven set at 35 °C, and a C18 reversed-phase column (250 mm × 4.6 mm, 5 μm, Agilent Technologies, Palo Alto, California, USA) for component separation. The mobile phase was composed of methanol/water (70:30, v/v) at a flow rate of 1.0 mL min^{-1}.

3. Results and Discussion

3.1. *Construction of Core–Shell Fe_3O_4@ZIF-8@MIP*

The strategy for fabricating MMOFs@MIPs (Fe_3O_4@ZIF-8@MIP) is schematically depicted in Figure 1. As a prototypical imidazole-based MOF with outstanding chemical and thermal stability as well as a high porosity and an easy preparation, ZIF-8 was selected as an ideal matrix to develop MMOFs. Carboxylate Fe_3O_4 particles were first prepared as a magnetic unit module through a simple solvothermal reaction using trisodium citrate as a stabilizer to obtain a negatively charged Fe_3O_4 surface, which favors the attachment of Zn^{2+} cations based on electrostatic interactions to initiate nucleation and growth to produce a ZIF-8 nanocrystal layer. As a result, magnetic Fe_3O_4@ZIF-8 was obtained and further employed as a support to fabricate Fe_3O_4@ZIF-8@MIP through a molecularly imprinted surface sol–gel process.

3.2. *Characterization of Fe_3O_4@ZIF-8@MIP*

3.2.1. SEM and TEM Characterization

The morphologies of the Fe_3O_4, Fe_3O_4@ZIF-8, and Fe_3O_4@ZIF-8@MIP particles were observed by SEM and TEM (Figure 1). As shown in Figure 1A–C, Fe_3O_4 (A), Fe_3O_4@ZIF-8 (B), and Fe_3O_4@ZIF-8@MIP (C) all exhibited nearly spherical shapes with narrow size distributions. In contrast to Fe_3O_4, Fe_3O_4@ZIF-8 had a rougher surface and a larger particle size. The TEM image (Figure 1D) revealed that the Fe_3O_4 nanoparticles were highly monodisperse with an average diameter of 200 nm. Moreover, many closely spaced cubic ZIF-8 crystals were visible on the surface of the Fe_3O_4 (Figure 1E), providing a large specific surface area for loading with the MIP layer, and the final product Fe_3O_4@ZIF-8@MIP (Figure 1F) had the same core–shell structure as Fe_3O_4@ZIF-8. To further confirm the successful synthesis of the Fe_3O_4@ZIF-8@MIP, the elemental distribution was also investigated (Figure S1). The image captured from several adhesive Fe_3O_4@ZIF-8@MIP particles exhibited a beautiful butterfly shape, indicating that Fe_3O_4@ZIF-8@MIP is composed of Fe, N, O, C, Si, and Zn, whereby Fe is mainly concentrated on the inside, in contrast to Si and Zn. The detected Zn, N, and Si signals also demonstrated the presence of ZIF-8 and MIP that was deposited on the entire surface of Fe_3O_4@ZIF-8.

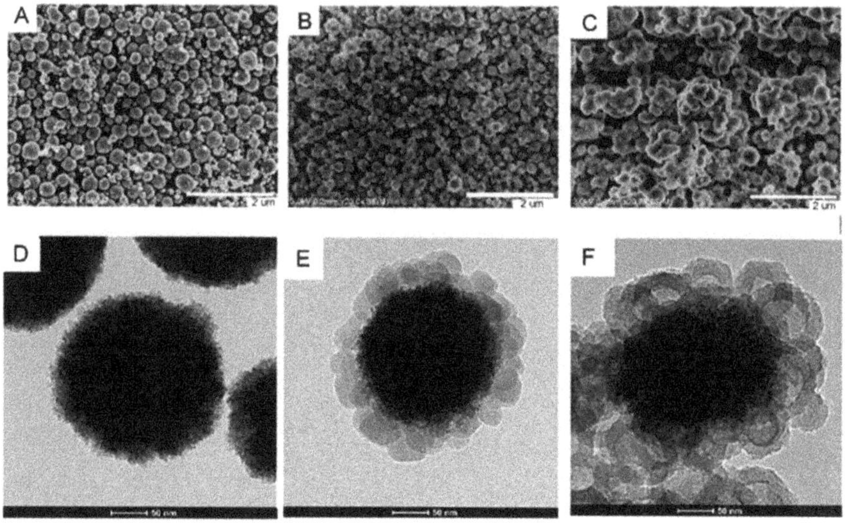

Figure 1. SEM images of Fe_3O_4 (**A**), Fe_3O_4@ZIF-8 (**B**), and Fe_3O_4@ZIF-8@MIP (**C**); TEM images of Fe_3O_4 (**D**), Fe_3O_4@ZIF-8 (**E**), and Fe_3O_4@ZIF-8@MIP (**F**).

3.2.2. FT−IR Measurements

As shown in Figure 2A, the functional groups of Fe_3O_4 (a), Fe_3O_4@ZIF-8 (b), Fe_3O_4@ZIF-8@MIP (c), and Fe_3O_4@ZIF-8@NIP (d) were analyzed by FT-IR spectra. The strong absorption band at 546 cm^{-1} originated from the Fe−O stretching vibration, and the peaks centered at 3353, 1593, and 1325 cm^{-1} correspond to the characteristic absorption of the −COOH group of trisodium citrate (Figure 2(Aa)), which indicates the successful synthesis of carboxylate Fe_3O_4. In Figure 2(Ab), in addition to the characteristic peaks of carboxylate Fe_3O_4, the new peaks appearing at around 1442 cm^{-1} and 900–1310 cm^{-1} are attributed to the tensile vibrations and in-plane bending of imidazole rings [23,24], and the absorption peak at 420 cm^{-1} is ascribed to Zn−N stretching vibration, which demonstrates the formation of ZIF-8 on the surface of Fe_3O_4. In addition, the characteristic peaks of Fe_3O_4@ZIF-8@MIP (Figure 2(Ac)) and Fe_3O_4@ZIF-8@NIP (Figure 2(Ad)) were almost the same. For example, the broad and intense absorption peak at 1033 cm^{-1} belongs to the Si−OH stretching vibration, and the peaks nearby at 1558 cm^{-1} and 1417 cm^{-1} are assigned to the stretching and the flexural vibrations of the N-H and $-CH_3$ groups [25], indicating the presence of aminopropyl groups. These spectra all confirm the successful encapsulation of Fe_3O_4@ZIF-8 by the MIPS via the siloxane copolymerization of APTES and TEOS. At the same time, the similarity of the peaks between Fe_3O_4@ZIF-8@MIP (Figure 2(Ac)) and Fe_3O_4@ZIF-8@NIP (Figure 2(Ad)) indicates the effective removal of the template-BPA, leaving abundant cavities and accessible sites for BPA adsorption.

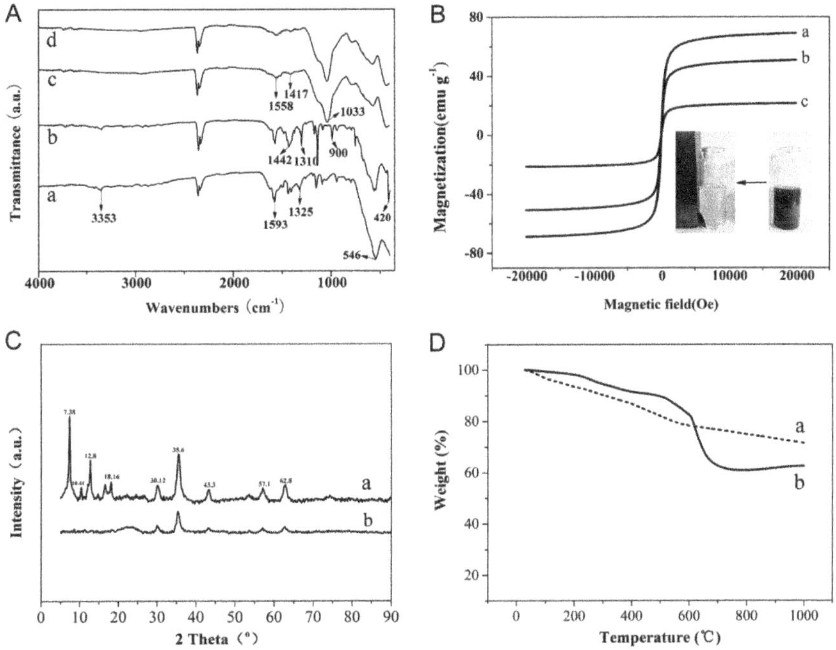

Figure 2. (**A**) FT−IR spectra of Fe_3O_4 (a), Fe_3O_4@ZIF-8 (b), Fe_3O_4@ZIF-8@MIP (c), and Fe_3O_4@ZIF-8@NIP (d); (**B**) magnetization curves of Fe_3O_4 (a), Fe_3O_4@ZIF-8 (b), and Fe_3O_4@ZIF-8@MIP (c); insert is a photograph of Fe_3O_4@ZIF-8@MIP suspended in solution before and after applying a magnetic field; (**C**) XRD patterns of Fe_3O_4@ZIF-8 (a), Fe_3O_4@ZIF-8@MIP (b); (**D**) TGA curves of Fe_3O_4@ZIF-8@MIP (a) and Fe_3O_4@ZIF-8 (b) in air.

3.2.3. VSM Analysis

The field-dependent magnetization curves with no hysteresis loop in Figure 2B show the supermagnetic features of Fe_3O_4 (a), Fe_3O_4@ZIF-8 (b), and Fe_3O_4@ZIF-8@MIP (c) with the determined saturation magnetization values of 69.02, 50.73, and 21.36 emu g^{-1}, respectively. The significant decrease in the saturation magnetization values was obviously caused by the layer of ZIF-8 and ZIF-8@MIPs, indicating the successful combination of ZIF-8 or ZIF-8@MIPs and magnetic naparticles. The resulting Fe_3O_4@ZIF-8@MIPs still possessed enough magnetic capacity for satisfactory separation. As evident in the insert image, the magnetic Fe_3O_4@ZIF-8@MIP can easily form a stable dispersion in ethanol solution without visible sedimentation; can be rapidly separated from the dispersion within a few seconds when a magnetic field is applied; and in turn can be redistributed in solution by simple shaking after the magnetic field is removed, which provides an accessible route for use as an absorbent in target separation or pollutant removal.

3.2.4. XRD Analysis

The crystal structures of synthesized Fe_3O_4@ZIF-8 (a) and Fe_3O_4@ZIF-8@MIP (b) were confirmed by powder XRD analysis (Figure 2C). The diffraction peaks of Fe_3O_4@ZIF-8 displayed at $2\theta = 7.38°$, 10.44°, 12.8°, and 18.16° are attributed to the (011), (002), (112), and (222) planes, which agrees well with the pure phase of ZIF-8, and the reflection peaks located at $2\theta = 30.12°$, 35.6°, 43.3°, and 30.12° correspond to the (220), (311), (400), (511), and (440) planes, in tune with the Fe_3O_4 lattice, which attests that the growth of ZIF-8 on Fe_3O_4 does not impact their respective crystalline integrities. Compared to Fe_3O_4@ZIF-8 (a), the characteristic diffraction peaks belonging to ZIF-8 were not observed and a new broad peak at 19.04–26.56° appeared in the pattern of the Fe_3O_4@ZIF-8@MIPs, which can be ascribed to the coverage effect of MIPs on Fe_3O_4@ZIF-8.

3.2.5. TG Analysis

TG analysis was performed to assess the thermal behavior of Fe_3O_4@ZIF-8@MIP (a) and Fe_3O_4@ZIF-8 (b), as shown in Figure 2D. When heated at temperatures of 25–500 °C, Fe_3O_4@ZIF-8 exhibited a gradual weight loss of 18%, resulting from the evaporation of the residual solvent and the pyrolysis of the carboxyl and silicon hydroxyl groups on the surface of ZIF-8 and Fe_3O_4, which was much higher than that of Fe_3O_4@ZIF-8 (10%) because the MIPs introduced more oxygen-containing groups. When further increasing the temperature to 700 °C, a sharp weight loss (27%) caused by the decomposition of the ZIF-8 framework occurred in Fe_3O_4@ZIF-8, whereas only a slight weight loss of 5.4% was observed in Fe_3O_4@ZIF-8@MIP, which can be ascribed to the protective effect of MIPs for the ZIF-8 framework. The high decomposition temperature reflects the excellent thermal stability of the Fe_3O_4@ZIF-8@ composite.

3.2.6. BET Measurements

The specific surface area and pore characteristics of Fe_3O_4@ZIF-8 and Fe_3O_4@ZIF-8@MIP have been estimated by N_2 adsorption and desorption experiments. As shown in Figure S2, a typical IV isotherm with a distinct hysteresis loop was observed for both the Fe_3O_4@ZIF-8 and Fe_3O_4@ZIF-8@MIPs, indicating their porous nature. Compared with Fe_3O_4@ZIF-8 (677.42 m^2/g), the BET surface area of the Fe_3O_4@ZIF-8@ significantly decreased to 4.21 m^2/g because of the encapsulation of Fe_3O_4@ZIF-8 by the MIPs. Using the BJH method, the total pore volume and average pore size was calculated to be 0.047 cm^3 g^{-1} and 35.98 nm for the Fe_3O_4@ZIF-8, as well as 0.036 cm^3 g^{-1} and 19.48 nm for the Fe_3O_4@ZIF-8@MIP.

3.3. *Adsorption Behaviors of MMOFs@MIP for BPA*

3.3.1. Adsorption Kinetic Experiment

The adsorption kinetics of Fe_3O_4@ZIF-8@MIPs, Fe_3O_4@ZIF-8@NIPs, Fe_3O_4@MIPs, and Fe_3O_4@NIPs for BPA with an initial concentration of 50 mg L^{-1} were also measured. As shown in Figure 3A,B, benefiting from the presence of imprinted cavities and recognition sites, Fe_3O_4@ZIF-8@MIPs not only exhibited a higher adsorption capacity for BPA compared to Fe_3O_4@ZIF-8@NIPs (Figure 3A) but also, more importantly, showed a remarkably increased mass transfer rate and faster binding kinetics with a threefold shorter adsorption equilibrium time than Fe_3O_4@MIPs (30 vs. 90 min, Figure 3B). This can be ascribed to the pores and specific surface area provided by the MOF.

Additionally, the kinetic data were further analyzed using pseudo-first-order and pseudo-second-order kinetic models. The results shown in Figure S3 illustrate that the dynamic BPA-adsorption behavior of Fe_3O_4@ZIF-8@MIPs and Fe_3O_4@ZIF-8@NIPs better fit a pseudo-second-order rate equation (R^2 = 0.9963, 0.9844) rather than a pseudo-first-order equation (R^2 = 0.9552, 0.9707), indicating that the adsorption process is controlled by the joint action of the solid MOF@polymer/liquid BPA solution interface rather than simple diffusion.

3.3.2. Equilibrium Binding Experiment

The static adsorption behavior of Fe_3O_4@ZIF-8@MIP and Fe_3O_4@ZIF-8@NIP was studied at room temperature with different initial concentrations of BPA (25–200 mg L^{-1}), as shown in Figure 3C. As a control, Fe_3O_4@MIP and Fe_3O_4@NIP were also employed (Figure S4). The adsorption capacity of Fe_3O_4@ZIF-8@MIP and Fe_3O_4@MIP was found to display an obvious concentration dependence. When incubated with 200 mg L^{-1} of BPA, the maximum Qe of 10.1 mg g^{-1} was reached for Fe_3O_4@ZIF-8@MIP, which is higher than the 3.3 mg g^{-1} for Fe_3O_4@ZIF-8@NIP based on nonspecific adsorption (Figure 3D). At the same time, due to the absence of MOFs as carriers, the adsorption capacity of traditional magnetic polymer-Fe_3O_4@MIP (Figure S4) was far lower than that of the Fe_3O_4@ZIF-8@MIP but slightly higher than that of the Fe_3O_4@ZIF-8@NIP, suggesting that there are synergetic effects of MOFs and MIPs that improve the adsorption capacity. The desirable adsorption characteristics make Fe_3O_4@ZIF-8@MIP an ideal candidate for the development of highly sensitive and selective adsorption and separation materials.

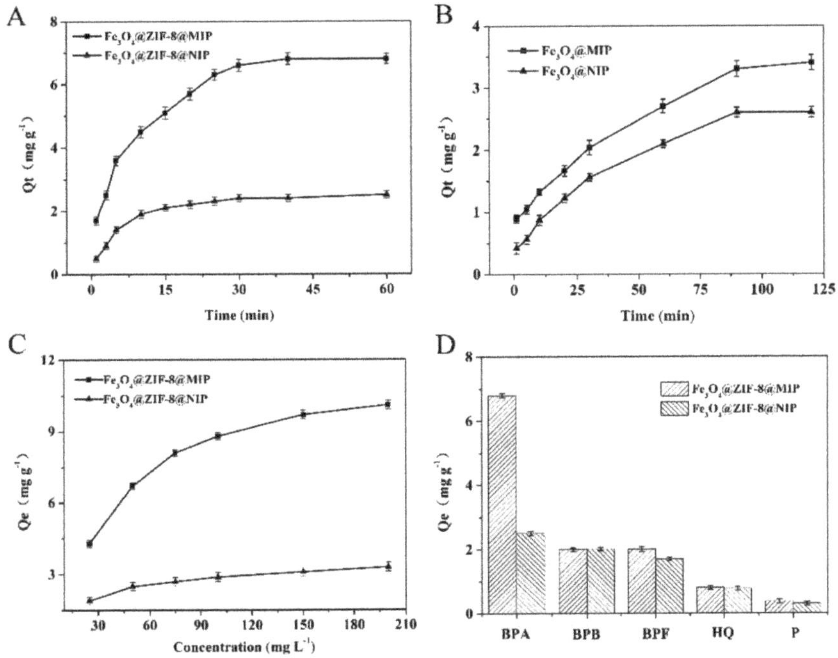

Figure 3. (**A**) Adsorption of BPA (50 mg L^{-1}) by Fe_3O_4@ZIF-8@MIP and Fe_3O_4@ ZIF-8@NIP at different incubation times (1, 3, 5, 10, 20, 25, 30, 40, and 60 min); (**B**) adsorption of BPA (50 mg L^{-1}) by Fe_3O_4@MIP and Fe_3O_4@NIP at different incubation times (1, 5, 10, 20, 30, 60, 90, and 120 min); (**C**) adsorption of BPA by Fe_3O_4@ZIF-8@MIP and Fe_3O_4@ZIF-8@NIP at different initial BPA concentrations (25, 50, 75, 100, 150, and 200 mg L^{-1}); (**D**) absorption of 50 mg L^{-1} initial BPA, as well as its structural analogues BPB, BPF, HQ, and P by Fe_3O_4@ZIF-8@MIP and Fe_3O_4@ZIF-8@NIP.

3.3.3. The Selectivity Evaluation for BPA

The selectivity of BPA adsorption by Fe_3O_4@ZIF-8@MIP and Fe_3O_4@ZIF-8@NIP was investigated. Phenol-containing compounds including BPB, BPF, HQ, and P with the same initial concentration of 50 mg L^{-1} were selected to test the target binding of Fe_3O_4@ZIF-8@MIP toward BPA (see Figure S5 for the corresponding chemical structures). Among them, BPA, BPB, and BPF have similar sizes and arrangements of phenol moieties but increasing hydrophobicity due to substitution at the bridging carbon atom; P, BP, and BPA possess varying numbers and arrangements of phenol groups and aromatic rings. The results are shown in Figure 3D, where the most favorable binding kinetics were observed for imprint molecule-BPA, with a much higher imprinted factor of 2.72 compared to its structural analogues, with values of 1.0 for BPB, 1.18 for BP, 1.05 for HQ, and 1.23 for P. This selectivity can be attributed to the imprinting process that left more specific cavities and recognition sites for the BPA template. In addition, Fe_3O_4@ZIF-8@NIP favored the adsorption of BPA, BPB, and BPF rather than the smaller analytes HQ and P, suggesting that silica alone does have some affinity for bisphenol species and that the adsorption process was controlled by the comprehensive effect of the ZIF-8 and MIP layers.

3.4. MSPE of BPA Using Fe_3O_4@ZIF-8@MIP

The Fe_3O_4@ZIF-8@MIP was employed as a separation tool for the enrichment of trace BPA via a magnetic solid phase extraction, and its application feasibility was studied. As an important parameter, the enrichment efficiency of Fe_3O_4@ZIF-8@MIP in different volumes (50, 100, 150, 200, and 250 mL) of solution containing 10 nmol BPA was investigated as shown in Figure 4A. High recovery rates of 88.1–96.5% were obtained, which implies that Fe_3O_4@ZIF-8@MIP has good enrichment capabilities for BPA at trace levels. In addition, the enrichment efficiencies of Fe_3O_4@ZIF-8@MIP and Fe_3O_4@ZIF-8@NIP were also evaluated using 0.5 mg L^{-1} BPA. As shown in Figure 5, the peak intensity after enrichment by Fe_3O_4@ZIF-8@MIP (a) was obviously greater than that of Fe_3O_4@ZIF-8@NIP (b), indicating the excellent enrichment selectivity of Fe_3O_4@ZIF-8@MIP for BPA.

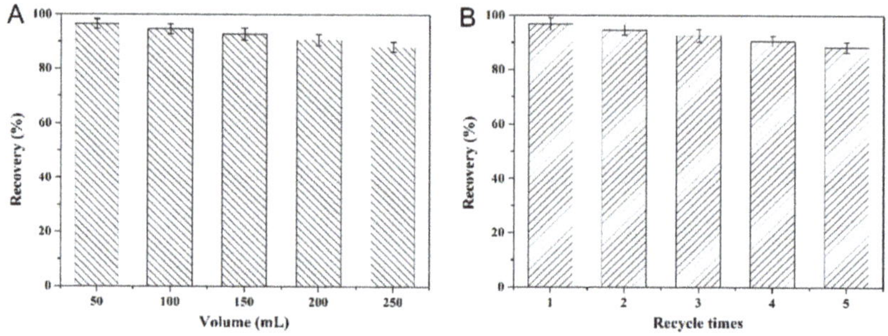

Figure 4. (**A**) Recovery rates of BPA in different sample volumes; (**B**) reusability of the Fe_3O_4@ZIF-8@MIP for BPA adsorption.

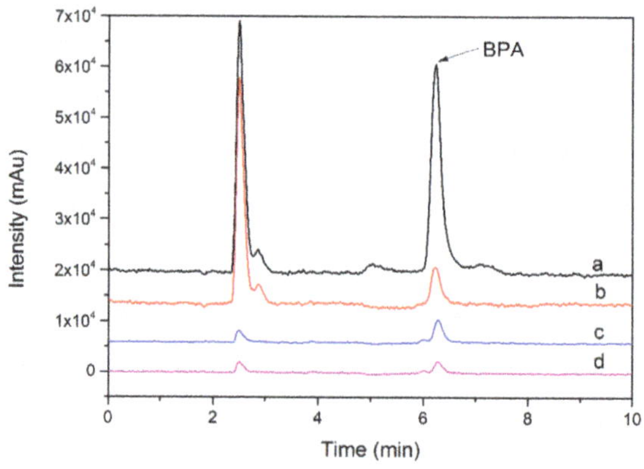

Figure 5. HPLC chromatograms of BPA after enrichment by Fe_3O_4@ZIF-8@MIP (a), Fe_3O_4@ZIF-8@NIP (b), and its presence in tap water (c) and bottled drinking water (d).

Reusability is another highly desired feature for commercial applications, so the reusability of Fe_3O_4@ZIF-8@MIP for 50 μg L^{-1} BPA was evaluated in five consecutive adsorption–desorption cycles, using methanol and acetic acid (3 mL, v/v = 1:9) as the regeneration agent. As depicted in Figure 4B, after five reuse cycles, the absolute recovery descended to 88% but remained above 90% for the first use, demonstrating its potential recyclability.

3.5. MSPE-HPLC for BPA Analysis in Real Samples

Using Fe_3O_4@ZIF-8@MIP as an absorbent in MSPE combined with widely accessible HPLC analysis, a highly sensitive and selective MSPE-HPLC detection method for BPA has been proposed with a wide linear range of 0.5–5000 μg L^{-1} and a low LOD of 0.1 μg L^{-1} (S/N = 3).

In order to evaluate its reliability and practicability, the developed MSPE-HPLC method based on the Fe_3O_4@ZIF-8@MIP was used to extract and determine BPA in real samples including tap water, mineral water, bottled drinking water, lemon juice, and canned hawthorn, among which tap water (Figure 5c) and bottled drinking water (Figure 5d) had detectable BPA levels of 7.5 ± 0.04 μg L^{-1} and 1.2 ± 0.02 μg L^{-1}, respectively. Compared to the glass packaging sample (canned hawthorn), BPA in the plastic packaging samples has an increased probability of being detected. It is worth noting that tap water was also contaminated by BPA at a higher concentration than in bottled drinking water, and the extent of this pollution requires more detailed research to be revealed.

To further verify the feasibility of this method, a recovery experiment was carried out by adding a standard solution of BPA at three concentration levels of 0.5, 1.0, and 10 μg L^{-1} into blank lemon juice, canned hawthorn, and mineral water samples. As shown in Table 1, the satisfactory recoveries ranging from 88.3% to 92.3% with an RSD of less than 3.6% were obtained, indicating the acceptable reliability and usability of this method.

Table 1. Recovery rates of BPA obtained from spiked real samples using the developed MSPE-HPLC method.

Sample	Spiked (μg L^{-1}, n = 3)	Recovery (%, n = 3)	RSD (%)
Lemon juice beverage	0.5	88.5	3.0
	1.0	89.3	2.3
	10.0	89.8	2.1
Canned Hawthorn	0.5	89.3	2.9
	1.0	88.3	3.3
	10.0	89.5	1.9
Mineral water	0.5	91.3	2.9
	1.0	92.3	3.6
	10.0	91.9	2.2

As shown in Table S1, compared with the previously developed HPLC methods [26–28] for BPA detection, our proposed method based on Fe_3O_4@ZIF-8@MIP exhibited higher sensitivity (lower LOD), broader applications including the applicability for simple water matrices as well as a complex fruit juice or canned fruit matrices (which is attributed to the composite effects and high selectivity of MIPs), accessible pores and a good accumulation of MOFs, and the efficient separation of magnetic Fe_3O_4.

4. Conclusions

We reported a simple and efficient layer-by-layer strategy for the synthesis of core–shell-structured Fe_3O_4@ZIF-8@MIPs without any special requirements for surface modification. The synthesized Fe_3O_4@ZIF-8@MIP integrates the desirable features of Fe_3O_4, MOFs, and MIPs and thus exhibits a strong magnetic responsiveness, an outstanding porosity, and a satisfactory adsorption selectivity for its target. Combined with HPLC analysis, Fe_3O_4@ZIF-8@MIPs can be successfully employed for the extraction and determination of BPA in a wide concentration range (0.5–5000 μg L^{-1}) with a low LOD (0.1 μg L^{-1}). We proposed a feasible strategy for the construction of multifunctional adsorption materials and illustrate the potential for the tailored application of MOFs in more fields.

Supplementary Materials: The following supporting information can be downloaded at https://www.mdpi.com/article/10.3390/foods11101408/s1, Figure S1: Elemental distribution maps in Fe_3O_4@ZIF-8@MIP; Figure S2: N_2 adsorption –desorption isotherms of Fe_3O_4@ZIF-8 (A) andFe_3O_4 @ZIF-8@MIP (B); Figure S3: Pseudo-first-order (A) and pseudo-second-order (B) absorption kinetic linear fitting curves of Fe_3O_4@ZIF-8@MIP and Fe_3O_4@ZIF-8@NIP; Figure S4: Adsorption capacity of

Fe_3O_4@MIP and Fe_3O_4@NIP towards BPA in different initial concentrations (25, 50, 100, 150, and 200 mg L^{-1}); Figure S5: Chemical structures of bisphenol A and its analogues used in this study; Table S1: Comparison of the proposed MSPE-HPLC method based on Fe_3O_4@ZIF-8@MIP with previously reported methods [29–31].

Author Contributions: Q.Z.: Writing—original draft, Formal Analysis, Visualization; H.W.: Methodology, Software, Investigation, Formal Analysis; Y.Z.: Data Curation, Validation, Project Administration; Z.X.: Conceptualization, Supervision; L.X.: Conceptualization, Methodology, Funding Acquisition, Resources, Supervision, Writing—Review and Editing. All authors have read and agreed to the published version of the manuscript.

Funding: This work was supported by the National Natural Science Foundation of China (31701684), the Incubation Program for the Youth Innovation Team in Shandong Province, and the Shandong Provincial Natural Science Foundation (ZR2017BC014).

Institutional Review Board Statement: Not applicable.

Informed Consent Statement: Not applicable.

Data Availability Statement: The datasets generated for this study are available upon request to the corresponding author. The data are not publicly available due to privacy.

Conflicts of Interest: The authors declare no conflict of interest.

References

1. Vilarinho, F.; Sendon, R.; Van der Kellen, A.; Vaz, M.F.; Sanches Silva, A. Bisphenol A in food as a result of its migration from food packaging. *Trends Food Sci. Technol.* **2019**, *91*, 33–65. [CrossRef]
2. Michałowicz, J. Bisphenol A-Sources, toxicity and biotransformation. *Environ. Toxicol. Pharmacol.* **2014**, *37*, 738–758. [CrossRef] [PubMed]
3. Suzuki, T.; Nakagawa, Y.; Takano, I.; Yaguchi, K.; Yasuda, K. Environmental fate of bisphenol A and its biological metabolites in river water and their xeno-estrogenic activity. *Environ. Sci. Technol.* **2004**, *38*, 2389–2396. [CrossRef] [PubMed]
4. Ricco, R.; Malfatti, L.; Takahashi, M.; Hill, A.J.; Falcaro, P. Applications of magnetic metal-organic framework composites. *J. Mater. Chem. A* **2013**, *1*, 13033–13045. [CrossRef]
5. Huo, S.H.; Yan, X.P. Facile magnetization of metal-organic framework MIL-101 for magnetic solid-phase extraction of poycyclic aromatic hydrocarbons in environmental water samples. *Analyst* **2012**, *137*, 3445–3451. [CrossRef]
6. Hu, Y.L.; Huang, Z.L.; Liao, J.; Li, G.K. A chemical bonding approach for fabrication of hybrid magnetic metal-organic framework-5: High efficient adsorbents for magnetic enrichment of trace analytes. *Anal. Chem.* **2013**, *85*, 6885–6893. [CrossRef]
7. Wu, Y.H.; Li, B.Y.; Wang, X.X.; Yu, S.J.; Pang, H.W.; Liu, Y.; Liu, X.Y.; Wang, X.K. Magnetic metal-organic frameworks (Fe_3O_4@ZIF-8) composites for U (VI) and Eu (III) elimination: Simultaneously achieve favorable stability and functionality. *Chem. Eng. J.* **2019**, *378*, 122105–122116. [CrossRef]
8. Alqadami, A.A.; Naushad, M.; Alothman, Z.A.; Ghfar, A.A. Novel metal-organic framework (MOF) based composite material for the sequestration of U (VI) and Th (IV) metal ions from aqueous environment. *ACS Appl. Mater. Interfaces* **2017**, *9*, 36026–36037. [CrossRef]
9. Chen, Y.; Xiong, Z.C.; Peng, L.; Gan, Y.Y.; Zhao, Y.M.; Shen, J.; Qan, J.H.; Zhang, L.Y.; Zhang, W.B. Facile preparation of core-shell magnetic metal-organic framework nanoparticles for the selective capture of phosphopeptides. *ACS Appl. Mater. Interfaces* **2015**, *7*, 16338–16347. [CrossRef]
10. Xiao, R.L.; Pan, Y.N.; Li, J.; Zhang, L.Y.; Zhang, W.B. Layer-by-layer assembled magnetic bimetallic metal-organic framework composite for global phosphopeptide enrichment. *J. Chromatogr. A* **2019**, *1601*, 45–52. [CrossRef]
11. Ke, F.; Jiang, J.; Li, Y.Z.; Liang, J.; Wan, X.C.; Ko, S. Highly selective removal of Hg^{2+} and Pb^{2+} by thiol-functionalized Fe_3O_4@metal-organic framework core-shell magnetic microspheres. *Appl. Surf. Sci.* **2017**, *413*, 266–274. [CrossRef]
12. Liu, G.Y.; Li, L.Y.; Gao, Y.H.; Gao, M.K.; Huang, X.D.; Lv, J.; Xu, D.H. A beta-cyclodextrin-functionalized magnetic metal organic framework for efficient extraction and determination of prochloraz and triazole fungicides in vegetables samples. *Ecotoxicol. Environ. Saf.* **2019**, *183*, 109546. [CrossRef]
13. Liu, Y.; Zhong, G.; Liu, Z.; Meng, M.; Liu, F.; Ni, L. Facile synthesis of novel photoresponsive mesoporous molecularly imprinted polymers for photo-regulated selective separation of bisphenol A. *Chem. Eng. J.* **2016**, *296*, 437–446. [CrossRef]
14. Mao, Y.L.; Kang, H.Y.; Guo, Y.F.; Chen, S.T.; Wang, Z.X. Synthesis of surface imprinted polymer upon modified kaolinite and study on the selective adsorption of BPA. *Desalin. Water Treat.* **2016**, *57*, 3947–3956. [CrossRef]
15. Matsumoto, K.; Tiu, B.D.B.; Kawamura, A.; Advincula, R.C.; Miyata, T. QCM sensing of bisphenol A using molecularly imprinted hydrogel/conducting polymer matrix. *Polym. J.* **2016**, *48*, 525–532. [CrossRef]
16. Poliwoda, A.; Moscipan, M.; Wieczorek, P.P. Application of molecular imprinted polymers for selective solid phase extraction of bisphenol A. *Ecol. Chem. Eng. S.* **2016**, *23*, 651–664. [CrossRef]

17. Luo, L.J.; Tan, W.; Xong, H.B.; Barrow, C.J.; He, P.; Yang, W.R.; Wang, H.B. TiO_2@ phenyl-functionalized mesoporous silica for removal of bisphenol A from water. *Desalin. Water Treat.* **2017**, *72*, 182–189. [CrossRef]
18. Qian, K.; Fang, G.Z.; Wang, S. A novel core-shell molecularly imprinted polymer based on metal-organic frameworks as a matrix. *Chem. Commun.* **2011**, *47*, 10118–10120. [CrossRef]
19. Iskierko, Z.; Sharma, P.S.; Prochowicz, D.; Fronc, K.; D'Souza, F.; Toczydlowska, D.; Stefaniak, F.; Noworyta, K. Molecularly imprinted polymer (MIP) film with improved surface area developed by using metal-organic framework (MOF) for sensitive lipocalin (NGAL) determination. *ACS Appl. Mater. Interfaces* **2016**, *8*, 19860–19865. [CrossRef]
20. Liu, H.; Mu, L.; Chen, X.; Wang, J.; Wang, S.; Sun, B. Core-Shell metal-organic frameworks/molecularly imprinted nanoparticles as absorbents for the detection of pyrraline in milk and milk powder. *J. Agric. Food Chem.* **2017**, *65*, 986–992. [CrossRef]
21. Liu, J.; Sun, Z.K.; Deng, Y.H.; Zou, Y.; Li, C.Y.; Guo, X.H.; Xiong, L.Q.; Gao, Y.; Li, F.Y.; Zhao, D.Y. Highly water-dispersible biocompatible magnetite particles with low cytotoxicity stabilized by citrate groups. *Angew. Chem. Int. Ed. Engl.* **2009**, *48*, 5875–5879. [CrossRef]
22. Zheng, J.N.; Lin, Z.; Lin, G.; Yang, H.H.; Zhang, L. Preparation of magnetic metal-organic framework nanocomposites for highly specific separation of histidine-rich proteins. *J. Mater. Chem. B* **2015**, *3*, 2185–2191. [CrossRef]
23. Bui, T.T.; Nguyen, D.C.; Hua, S.H.; Chun, H.; Kim, Y.S. Sonochemical Preparation of a Magnet-Responsive Fe_3O_4@ ZIF-8 Adsorbent for Efficient Cu^{2+} Removal. *Nanomaterials* **2022**, *12*, 753–764. [CrossRef]
24. Xiong, Z.; Zheng, H.; Hu, Y.; Hu, X.; Ding, W.; Ma, J.; Li, Y. Selective adsorption of Congo red and Cu (II) from complex wastewater by core-shell structured magnetic carbon@ zeolitic imidazolate frameworks-8 nanocomposites. *Sep. Purif. Technol.* **2021**, *277*, 119053–119070. [CrossRef]
25. Rostkowska, H.; Szczepaniak, K.; Nowak, M.J.; Leszczynski, J.; KuBulat, K.; Person, W.B. Thiouracils. 2. Tautomerism and infrared spectra of thiouracils. Matrix-isolation and ab initio studies. *J. Am. Chem. Soc.* **1990**, *112*, 2147–2160. [CrossRef]
26. Liu, J.Z.; Wang, W.Z.; Xie, Y.F.; Huang, Y.Y.; Liu, Y.L.; Liu, X.J.; Zhao, R.; Liu, G.Q.; Chen, Y. A novel polychloromethylstyrene coated superparamagnetic surface molecularly imprinted core–shell nanoparticle for bisphenol A. *J. Mater. Chem.* **2011**, *21*, 9232–9238. [CrossRef]
27. Xu, Z.G.; Yang, Z.L.; Liu, Z.M. Development of dual-templates molecularly imprinted stir bar sorptive extraction and its application for the analysis of environmental estrogens in water and plastic samples. *J. Chromatogr. A* **2014**, *1358*, 52–59. [CrossRef]
28. Li, J.; Zhang, X.B.; Liu, Y.X.; Tong, H.W.; Xu, Y.P.; Liu, S.M. Preparation of a hollow porous molecularly imprinted polymer using tetrabromobisphenol A as a dummy template and its application as SPE sorbent for determination of bisphenol A in tap water. *Talanta* **2013**, *117*, 281–287. [CrossRef]
29. Liu, X.L.; Wang, Y.H.; Ren, S.Y.; Li, S.; Wang, Y.; Han, D.P.; Qin, K.; Peng, Y.; Han, T.; Gao, Z.X.; et al. Fabrication of magnetic Al-based Fe_3O_4@MIL-53 metal organic framework for capture of multi-pollutants residue in milk followed by HPLC-UV. *Molecules* **2022**, *27*, 2088–2098. [CrossRef]
30. Wu, X.; Li, Y.; Zhu, X.; He, C.; Wang, Q.; Liu, S. Dummy molecularly imprinted magnetic nanoparticles for dispersive solid-phase extraction and determination of bisphenol A in Water Samples and Orange Juice. *Talanta* **2017**, *162*, 57–64. [CrossRef] [PubMed]
31. Feng, X.; Xu, X.; Liu, Z.; Xue, S.; Zhang, L. Novel functionalized magnetic ionic liquid green separation technology coupled with high performance liquid chromatography: A rapid approach for determination of estrogens in milk and cosmetics. *Talanta* **2020**, *209*, 120542–120551. [CrossRef] [PubMed]

foods

MDPI

Article

A High-Luminescence Biomimetic Nanosensor Based on N, S-GQDs-Embedded Zinc-Based Metal–Organic Framework@Molecularly Imprinted Polymer for Sensitive Detection of Octopamine in Fermented Foods

Ying Guo, Guanqing Yuan, Xuelian Hu, Jinni Zhang and Guozhen Fang *

State Key Laboratory of Food Nutrition and Safety, Tianjin University of Science and Technology, Tianjin 300457, China; guoyingaxx@163.com (Y.G.); 15731366559@126.com (G.Y.); hxl523686206@126.com (X.H.); zjn939699@163.com (J.Z.)
* Correspondence: fangguozhen@tust.edu.cn; Tel.: +86-22-6091-2493; Fax: +86-22-6091-2493

Abstract: In this study, a novel fluorescent molecularly imprinted nanosensor (N, S-GQDs@ZIF-8@MIP) based on the nitrogen and sulfur co-doped graphene quantum dots decorated zeolitic imidazolate framework-8 was constructed for the detection of octopamine (OA). Herein, ZIF-8 with a large surface area was introduced as a supporter of the sensing system, which effectively shortened the response time of the sensor. Meanwhile, high green luminescent N, S-GQDs and a maximum emission wavelength of 520 nm under 460 nm excitation and a 12.5% quantum yield were modified on the surface of ZIF-8 as a signal tag that can convert the interactions between the sensor and OA into detectable fluorescent signals. Finally, N, S-GQDs@ZIF-8@MIP was acquired through the surface molecular imprinting method. Due to the synergy of N, S-GQDs, ZIF-8, and MIP, the obtained sensor not only demonstrated higher selectivity and sensitivity than N, S-GQDs@ZIF-8@NIP, but also displayed faster fluorescence response than N, S-GQDs@MIP. Under optimal conditions, the developed sensor presented a favorable linear relationship in the range of 0.1–10 mg L^{-1} with a detection limit of 0.062 mg L^{-1}. Additionally, the proposed N, S-GQDs@ZIF-8@MIP strategy was effectively applied to the detection of OA in fermented samples, and the obtained results had a satisfactory correlation with those of HPLC.

Keywords: N/S co-doped graphene quantum dots; zeolitic imidazolate framework-8; molecularly imprinted polymer; octopamine; fluorescent sensor

Citation: Guo, Y.; Yuan, G.; Hu, X.; Zhang, J.; Fang, G. A High-Luminescence Biomimetic Nanosensor Based on N, S-GQDs-Embedded Zinc-Based Metal–Organic Framework@Molecularly Imprinted Polymer for Sensitive Detection of Octopamine in Fermented Foods. *Foods* **2022**, *11*, 1348. https://doi.org/10.3390/foods11091348

Academic Editors: Antonello Santini and Arun K. Bhunia

Received: 13 February 2022
Accepted: 4 May 2022
Published: 6 May 2022

1. Introduction

Biogenic amines (BAs) are a type of biologically active containing nitrogen organic compound related to the physiological functions of humans [1]. Octopamine (OA) is a typical organic matter of BAs, named after its initial discovery in octopus saliva and has been known as a natural β3-adrenergic receptor agonist in the nervous system [2]. The intaking of excessive OA can cause toxic effects [3], such as headaches, vomiting, abnormal blood pressure, skin allergies, and other symptoms, and has led to brain hemorrhages in severe cases. In 2004, OA was banned in all sports competition since its stimulating properties may reduce the immunity of the organism and damage the sensory organs [4]. Therefore, as OA poses a potential public health threat, the detection of OA has become an important indicator of food safety. To date, various traditional analytical methods for the quantitative determination of OA have been developed, including electrochemical sensor detection [5], pseudo-ELISA [6], fast-scan cyclic voltammetry (FSCV) [7], LC-MS/MS method [8], and high-performance liquid chromatography (HPLC) [8]. Although these methods have made large progress for the detection of OA, when confronted with the detection of OA in fermented foods, these methods are limited by the need for laborious pretreatment, high costs, and interference by environment [9]. Therefore, it is particularly important to develop inexpensive and highly sensitive approaches for the detection of OA.

Graphene quantum dots (GQDs) have attracted widespread attention as a novel kind of nanomaterial due to their unique properties, such as distinctive electron transfer, strong chemical inertness, and low toxicity [10]. Recently, many studies have proved that the photoelectric characteristics of GQDs can be efficiently adjusted via doping heteroatoms [11]. Additionally, it was demonstrated that doped GQDs have excellent optical properties compared to non-doped GQDs [10,12]. The introduction of the N atom in the GQD structure helps to improve the quantum yield. Previous works have indicated that nitrogen-doped GQDs exhibit a higher quantum yield than non-doped GQDs [13]. S, as a heteroatom, is also widely used for GQD synthesis because it can effectively change the electron structure of the GQDs to produce multiple emission peaks by introducing S-related energy levels between the π and π^* of C [14]. Li et al. have implied that S-GQDs had a sensitive reaction to Fe^{3+} because the introduction of a S atom improved the electronic properties and surface chemical reactivities [15]. After doping these GQDs, the N, S-GQDs exhibited excellent optical performance, including multiple emission peaks and high fluorescent quantum yield. For example, Qu et al. developed N, S-GQDs with three absorption bands independent of the excitation wavelength [16]. Thus, N, S-GQDs can be considered fluorescent materials with good application prospects.

Metal–organic frameworks (MOFs) composed of central metal ions and organic ligands are also known as porous coordination polymers (PCPs) [17]. Over the past few years, MOFs have drawn increasing attention in the research areas of gas adsorption and separation [18], drug delivery [19], electrochemical energy storage [20], and sensing [21] due to their special large surface areas, tunable porosity, and versatile architectures [22]. However, to date, their sensing functions of MOFs have lacked signal transduction capacity, which has severely restricted their development. Encapsulating luminescent molecules or nanoparticles into the non-luminescent MOFs could solve this problem [23]. Our previous research has revealed that embedding CDs as the fluorescence signal into the zeolitic imidazolate framework (ZIF-8) can acquire sensitizing materials [24]. Therefore, this strategy can not only obtain composite material with fluorescence-sensing properties but also improve the dispersion of luminescent molecules. Although the as-prepared composite material has good sensitivity, it still faces interference from other substances in the detection system due to the non-specificity of the fluorescent sensor.

In recent years, the use of molecular imprinting technology to modify MOFs has gradually aroused the interest of researchers. Molecular imprinting technology is a biomimetic technique based on the specific reaction between the antigen and antibody, which has been applied to prepare the polymers with specific structures [25]. The formed polymers are named molecularly imprinted polymers (MIPs) and not only exhibit selectivity for a target molecule or a group of structurally related analogies, but also have higher stability and longer service lives as well as tolerance to different environments [26]. Based on these excellent properties, MIPs are widely used to build sensing, separation, and catalytic platforms [27]. In the past few years, there have been a few studies on octopamine (OA) determination based on molecularly imprinted fluorescent nanoparticles [28]. Although the aforementioned studies determined that molecular imprinting technology can be used as a facile, reliable, and rapid strategy for OA detection via the formation of specific recognition sites, such methods frequently have low sensitivity and selectivity, limiting their applications in the field of optical sensing. Hence, it is of great significance to develop a fluorescent sensor with exceptional sensitivity and selectivity.

Inspired by this previous research, we designed and developed a novel fluorescent nanosensor N, S-GQDs@ZIF-8@MIP based on the N, S-GQDs-embedded zeolitic imidazolate framework-8 @ molecularly imprinted polymer. In this work, as the supporter, ZIF-8 with a large surface area was first prepared via the self-assembly method. Subsequently, N, S-GQDs-functionalized ZIF-8 (N, S-GQDs@ZIF-8) was synthesized by encapsulating N, S-GQDs using extremely bright-green light within ZIF-8 under convenient room temperature stirring. Finally, N, S-GQDs@ZIF-8@MIP with high selectivity and recognition ability for target molecules was fabricated through the sol–gel approach in a mixed solution containing N, S-GQDs (fluorophore), ZIF-8 (supporter), OA (template), 3-aminopropyl triethoxysilane (functional monomer), and tetraethyl orthosilicate (cross-linker). The incorporated bright-green luminescent N, S-GQDs, as signal bands, converts the chemical reaction with OA into detectable fluorescence signals, thereby realizing a highly sensitive response to OA. ZIF-8 with a high specific surface area, as a supporter, can improve the dispersion of N, S-GQDs and increase the mass transfer rate. Additionally, the imprinted layer decorated on the surface of the N, S-GQDs@ZIF-8 provided the specific recognition site for the target molecule. The uniform morphology, superior photostability, highly selective recognition performance, and ultra-sensitive fluorescence response of the as-obtained N, S-GQDs@ZIF-8@MIP were highly confirmed by the characterizations. The practicality of the proposed sensor was verified through realizing sensitive detection of OA in real samples. The preparation and characterization of the obtained N, S-GQDs@ZIF-8@MIP were discussed in detail.

2. Materials and Methods

2.1. Materials and Instruments

All the reagents were analytically pure and used as received. 2-Methylimidazole (2-MI, 99%) and *N,N*-Dimethylformamide (DMF) were acquired from TCI (Shanghai, China). 3-aminopropyl triethoxysilane (APTES, 98%), citric acid (CA, 99%), tetraethylorthosilicate (TEOS, 99%), zinc nitrate hexahydrate ($Zn(NO_3)_2{\cdot}6H_2O$, 99%), thiourea, glucose and sucrose were obtained from Aladdin Chemistry (Shanghai, China). KCl, $MgCl_2$, NaCl, $FeCl_3$, L-Arginine (L-Arg), D-Tryptophan (D-Try), L-Methionine (L-Met), L-Cysteine (L-Cys), L-Histidine (L-His), L-Tryptophan (L-Try), L-Tyrosine (L-Tyr), Dopamine (DA) and OA were supplied by Shanghai Macklin Biochemical Co., Ltd. (Shanghai, China). $Na_2HPO_4{\cdot}12H_2O$ was purchased from Sinopharm Chemical Reagent Co., Ltd. (Tianjin, China). Information on the instruments was provided in the supporting information.

Fluorescence spectra were performed on an F-7100 fluorescence spectrophotometer (Thermo, Waltham, MA, USA) at the excitation wavelength of 460 nm. Fourier-transform infrared (FT-IR) was performed to analyze the obtained materials on a Tensor-37 FTIR spectrophotometer (Bruker, Bremen, Germany). Scanning electron microscope (SEM) and transmission electron microscope (TEM) images were used to show the surface and structure of the obtained materials (Waltham, MA, USA). Ultraviolet-visible (UV-Vis) absorption spectra were obtained on UV-Cary100 spectrophotometer (Shimadzu, Kyoto, Japan). Thermogravimetric analysis (TGA) was carried out with a PTC-10 A thermal gravimetric analyzer (Rigaku Corp., Tokyo, Japan) from room temperature to 900 °C. X-ray diffraction (XRD) data were measured at the angular range of 5–60 degrees (2θ) with Cu Kα radiation.

2.2. Synthesis of ZIF-8, N, S-GQDs and the N, S-GQDs@ZIF-8@MIP Composite

ZIF-8 was prepared as described in the previous literature via the one-pot synthesis method at room temperature [29]. In brief, 125 mg of $Zn(NO_3)_2{\cdot}6H_2O$ and 275.9 mg of 2-MI were dispersed in 10 mL of methanol with sonicating for 30 min. Subsequently, the solutions were transferred to a 100 mL flask and mixed well under a magnetic stirrer. This reaction was maintained for about 50 min. The resulting products were centrifuged at 10,000 rpm for 5 min and dried at 50 °C under a vacuum for 8 h.

N, S-GQDs were synthesized via the solvothermal route based on the method reported previously [16]. Briefly, 1 mmol citric acid and 3 mmol thiourea were added into 3 mL *N,N*-Dimethylformamide with magnetic stirring for 30 min. The mixture was then moved to an autoclave and maintained at 180 °C for 8 h under N_2 gas. The N, S-GQDs were collected from the suspension, centrifuged at 10,000 rpm for 10 min, and washed with ethanol. The precipitate was then dispersed into ethanol for subsequent use.

The surface-imprinted sol–gel technique was used to synthesize N, S-GQDs@ZIF-8@MIP. The product synthesis was achieved as follows. The as-prepared ZIF-8 and 200 μL N, S-GQDs were added into ethanol and stirred continuously until completely dispersed. Then, 1 mmol OA and 4 mmol APTES were dispersed into the solution with magnetic stirring for 30 min. Under continuous stirring, 6 mmol TEOS and ammonia water were added stepwise. After being pre-assembled, the mixed solution was sealed with a sealing film and stirred under a N_2 atmosphere for 10 h. The final material was obtained through washing with ethanol until no template molecules of OA could be detected. N, S-GQDs@ZIF-8@NIP was acquired using an identical procedure, but no OA was added during preparation. N, S-GQDs@MIP was synthesized in the same manner while avoiding the addition of ZIF-8.

2.3. Fluorescence Measurement

Fluorescence (FL) measurements were accomplished under an excitation wavelength of 460 nm using a Hitachi FL-7000 fluorescence spectrophotometer (Tokyo, Japan). In total, 1.0 mg N, S-GQDs@ZIF-8@MIP, N, S-GQDs@ZIF-8@NIP, or N, S-GQDs@MIP was dispersed in different 2 mL OA concentration solutions at 20 ± 5 °C, under laboratory relative humidity of 50–70%. After about 40 min of incubation, a quantitative analysis was performed. The fluorescence intensity of the sensing system at 520 nm was tested under excitation of 460 nm. Selectivity was evaluated by adding OA or other potentially coexisting substances under the above experimental conditions.

2.4. Pretreatment of Samples

In order to evaluate the accuracy of the developed N, S-GQDs@ZIF-8@MIP sensor in detecting OA in samples, wine and white vinegar were taken as representative fermented samples. Additionally, we studied three spiked concentration (1, 5, 9 mg L^{-1}) levels via spike recovery experiments. The samples were vortexed evenly and placed overnight. The procedure used to prepare the samples was as follows. In short, 4 mL of 5% trichloroacetic acid was mixed with 1 mL of the sample for 2 min. After centrifugation was performed at 3600 rpm for 10 min to collect the supernatant, 4 mL of n-hexane was added to the supernatant to remove the fat. Next, the extract's pH was adjusted to 7.0 with an NaOH solution, and the extract was evaporated to dryness. Finally, the residue was redissolved with 1 mL of ultrapure water and filtered with a microporous membrane for FL detection. Then, for FL analysis, 1.0 mg N, S-GQDs@ZIF-8@MIP was placed in a 1 mL sample solution and incubated for 40 min. This method was verified via HPLC. The pretreatment method for wine and white vinegar was provided in the supporting information.

For HPLC detection, the samples that were extracted and defatted as abovementioned were derivatized with dansyl chloride and then analyzed according to the National Standards of the People's Republic of China (GB 5009.208-2016). The analyses were performed using an HPLC system equipped with an ultraviolet detector. The HPLC operating parameters were as follows: The separation was carried out with a reversed phase column (C18, 250 × 4.6 mm, 5 μm particle size) with a flow rate of 0.8 mL min^{-1} and a column temperature of 35 °C. The injection volume was 20 μL, and the detection wavelength was 254 nm.

3. Results and Discussion

3.1. Preparation of N, S-GQDs, ZIF-8 and N, S-GQDs@ZIF-8@MIP

As displayed in Scheme 1, a novel synthetic strategy was proposed to prepare a fluorescent sensor constructed from MIP, N, S-GQDs, and ZIF-8 for the sensitive detection of OA. N, S-GQDs were synthesized through the solvothermal route based on citric acid as a carbon source and thiourea as nitrogen and sulfur sources (Scheme 1a). Compared to other GQDs, the obtained N, S-GQDs were excitation-dependent and showed different types of emission light under multiple excitation conditions [30]. It is worth noting that N, S-GQDs not only displayed extremely bright green light but also had a quantum yield (QY) as high as 12.5% [16]. The above phenomenon can be explained as follows. On the one hand, the high QY may be attributed to the doping of nitrogen atoms that changing the optical and electronic structure of the graphene quantum dots [31]. On the other hand, the green-fluorescent N, S-GQDs could benefit from the conjugation of the C=N, C=S, and C=O groups with the graphene core [32]. As presented in Scheme 1b, the high surface area ZIF-8 was prepared via the one-pot synthesis method. The introduction of ZIF-8 with a high specific surface area not only improved the dispersibility of GQDs, but also increased the mass transfer rate efficiently.

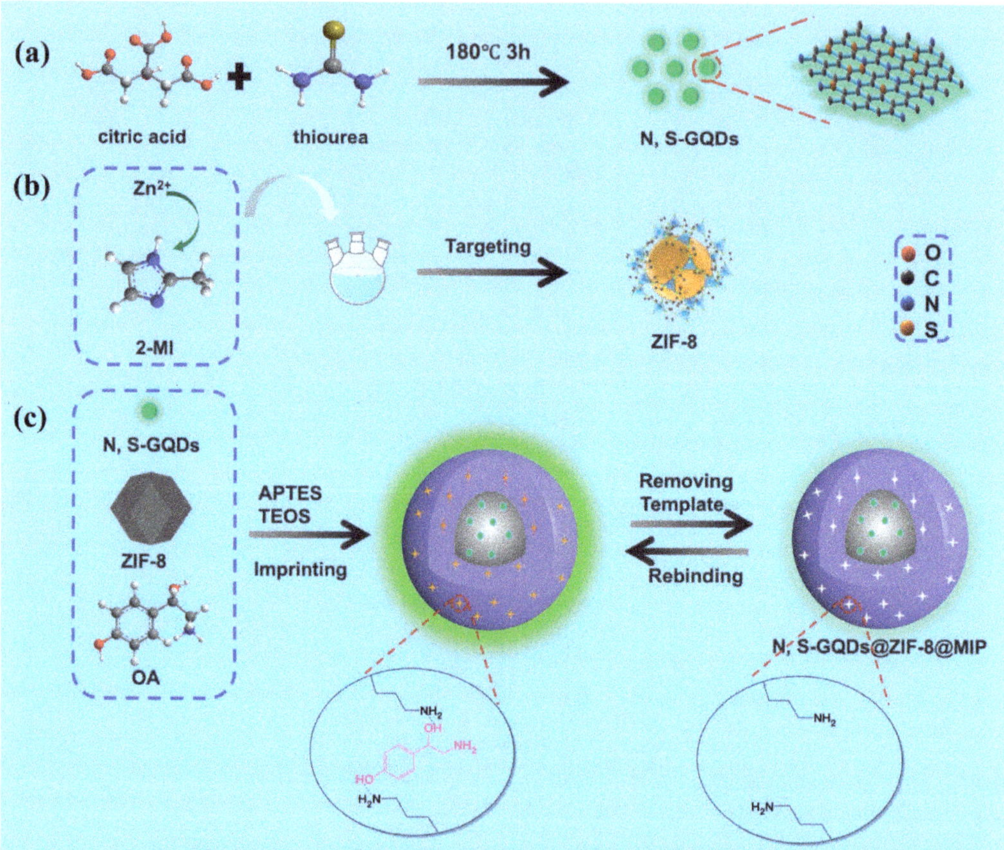

Scheme 1. Preparation process of N, S-GQDs (**a**), ZIF-8 (**b**), and N, S-GQDs@ZIF-8@MIP sensor (**c**).

As illustrated in Scheme 1c, N, S-GQDs@ZIF-8@MIP was constructed using the surface-imprinted sol–gel technique. The condensation reaction was conducted under the presence of OA, APTES, TEOS, and aqueous ammonia solution. During the polymerization process, the ultra-small size N, S-GQDs were first embedded in ZIF-8 to form N, S-GQDs@ZIF-8, and then the imprinted layer was decorated on the N, S-GQDs@ZIF-8 according to the physical deposition. After, the template molecule of OA was removed via ethanol, leaving behind 3D cavities in the imprinting layer, and obtaining a N, S-GQDs@ZIF-8@MIP sensor with selective binding sites fitting the shape of OA.

In the preparation processing of MIPs, the molar ratios of APTES and TEOS, the added quantity of supporter ZIF-8, and the amount of the fluorescent indicator N, S-GQDs had a certain impact on the imprinting factor (IF), which can directly reflect the polymer performance. Hence, to obtain materials with superior fluorescence and adsorption properties, we optimized the above aggregation conditions one by one, as presented in the supporting information file (Tables S1–S3).

3.2. Characterization of Materials

SEM and TEM were used to observe the morphology of the obtained ZIF-8, N, S-GQDs@ZIF-8@MIP (MIP), and N, S-GQDs@ZIF-8@NIP (NIP). Figure 1a shows that the as-prepared ZIF-8 conformed to its well-known dodecahedron shape. As displayed in Figure 1d, the unmodified ZIF-8 had a smooth surface and uniform particle size. After grafting, the surface morphology of MIP (Figure 1b) and NIP (Figure 1c) presented an obvious change, and an apparent porous structure could be observed. This phenomenon indicated that the polymers were formed. TEM images of N, S-GQDS@ZIF-8 are provided in the supporting information file (Figure S1).

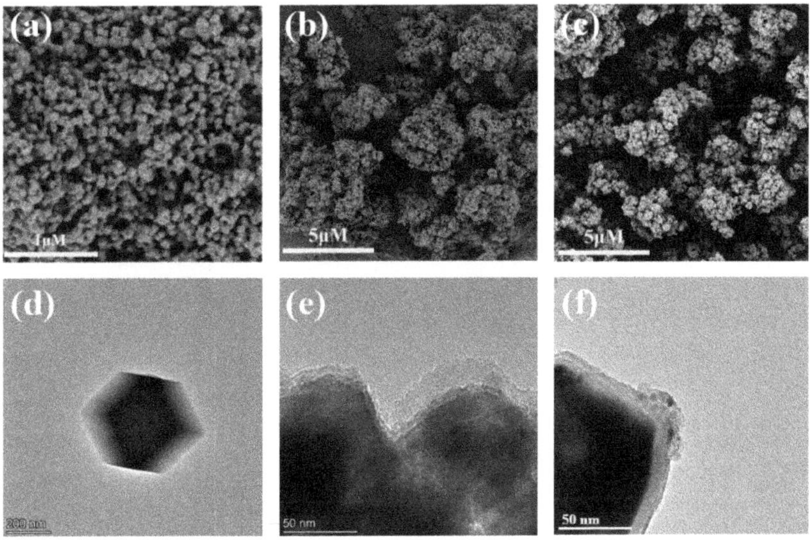

Figure 1. SEM images of ZIF-8 (**a**), N, S-GQDs@ZIF-8@MIP (**b**) and N, S-GQDs@ZIF-8@NIP (**c**); TEM images of ZIF-8 (**d**), N, S-GQDs@ZIF-8@MIP (**e**) and N, S-GQDs@ZIF-8@NIP (**f**).

The characteristics of functional groups of MIP (a), NIP (b), and ZIF-8 (c) were confirmed via FT-IR (Figure 2a). For curves a and b, characteristic peaks at 1185 cm^{-1} (C-S) and 782 cm^{-1} (C=S) can be seen in spectra a and b (Figure 2a), which indicated the successful introduction of N, S-GQDs into the composite materials [32,33]. The ZIF-8 presented characteristic absorption peaks at 1587 cm^{-1} (C=N), 1145 cm^{-1} (C-N) and 424 cm^{-1} (Zn-N bonds), as shown in Figure 2a(c). However, in curves a and b, it can be seen that the absorption peaks of ZIF-8 at 424 cm^{-1} and 1587 cm^{-1} were weakened, illustrating that the molecularly imprinted layer was successfully modified on the surface of ZIF-8. The two peaks at 1089 cm^{-1} (Si-O-Si) and 485 cm^{-1} (Si-O) indicate the existence of APTES and TEOS, respectively. The characteristic peaks in the range of 500–4000 cm^{-1} for MIP and NIP were almost identical, revealing that the template OA was removed from the imprinted polymers. The above findings indicate that MIP and NIP were successfully prepared.

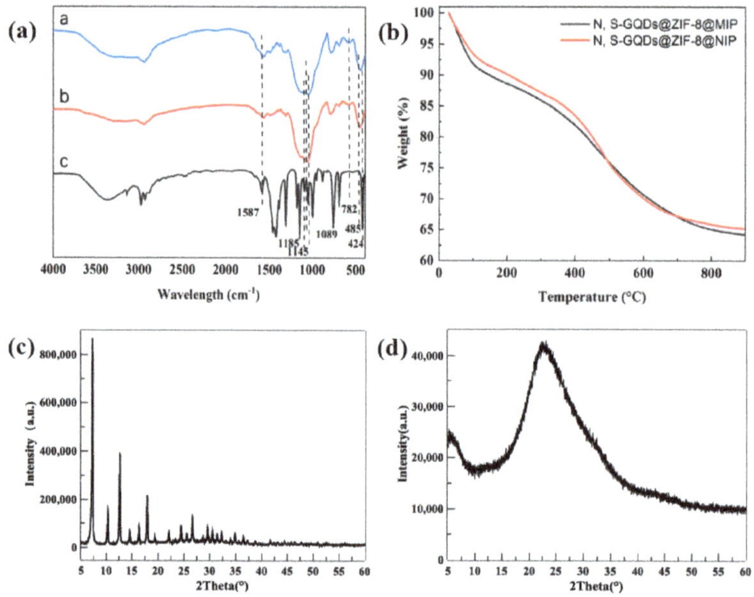

Figure 2. (**a**) FT-IR spectra of N, S-GQDs@ZIF-8@MIP-a, N, S-GQDs@ZIF-8@NIP-b, and ZIF-8-c. (**b**) TGA curves of N, S-GQDs@ZIF-8@MIP and N, S-GQDs@ZIF-8@NIP. XRD curves of ZIF-8 (**c**) and N, S-GQDs@ZIF-8@MIP (**d**).

TGA of MIP and NIP were performed under a stream of nitrogen. The results of weight as a function of temperature were depicted in Figure 2b. Obviously, there were three main stages of weight loss for MIP and NIP. The first stage of weight loss of MIP starting from 25 °C to 100 °C was approximately 8.50% due to the volatilization of moisture on the surface of the material and decomposition of the target that was not completely eluted. However, for NIP, weight loss was only due to volatilization of moisture on the surface of the material. Immediately afterwards, at 100–480 °C, the weight of the composite materials (MIP and NIP) was lost due to decomposition of the imprinting layer. The thermal gravimetric curves of MIP and NIP were almost the same due to the decomposition of the framework and organic polymer with the weight loss of 12.82%. The downward trends for the weight loss of MIP and NIP were similar to those of pure ZIF-8 crystals [34], demonstrating that the polymer was wrapped on the ZIF-8.

The crystal structure of synthesized ZIF-8 and MIP were analyzed using XRD. The pattern of ZIF-8 shown in Figure 2c matched well with the simulated results for the reported single-crystal data in the literature [35], revealing the successful synthesis of ZIF-8. The synthesized MIP showed the XRD pattern illustrated in Figure 2d. After polymerization, the diffraction peaks of ZIF-8 almost disappeared, which indicated the successful coating of MIP. Meanwhile, a new broad characteristic vibration peak located at $2\theta = 23.5°$ appeared in MIP (Figure 2d), which stemmed from the introduction of MIPs and N, S-GQDs to the (002) graphite planes.

3.3. Photoluminescence Properties and Kinetic Absorption of N, S-GQDs@ZIF-8@MIP

The optical properties of N, S-GQDS@ZIF-8@MIP were also investigated. As illustrated in Figure 3a, N, S-GQDS@ZIF-8@MIP exhibited symmetrical fluorescence emissions at 520 nm under excitation at 460 nm. The FL intensity of N, S-GQDS@ZIF-8@MIP was effectively enhanced when OA molecules were included in the imprinted cavities (Figure 3b). The FL intensity of N, S-GQDS@ZIF-8@MIP was recovered to 109.5% of that of N, S-GQDS@ZIF-8@NIP when removing the target molecules with ethanol as the elution solution.

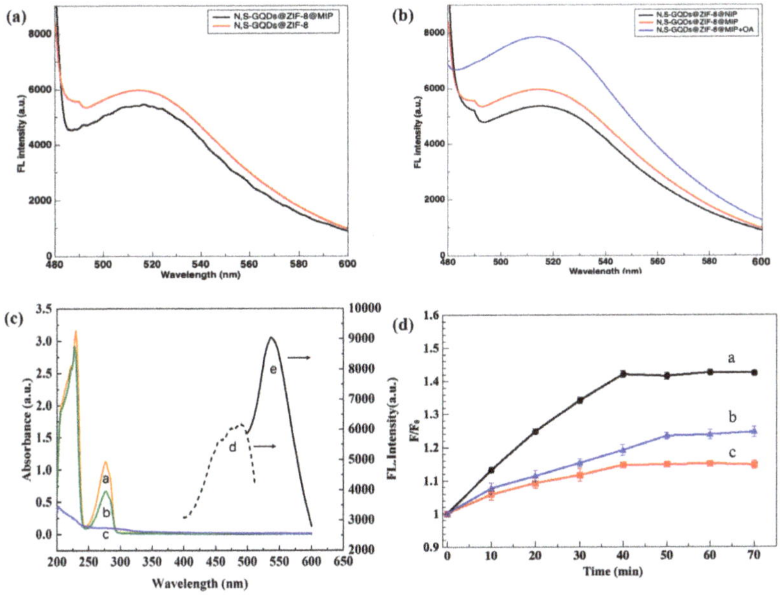

Figure 3. (**a**) Fluorescence spectra N, S-GQDs@ZIF-8 and N, S-GQDs@ZIF-8@MIP under 460 nm excitation. The concentrations of N, S-GQDs@ZIF-8 and N, S-GQDs@ZIF-8@MIP were 1 mg mL^{-1}. (**b**) Fluorescence emission spectra of N, S-GQDs@ZIF-8@MIP, N, S-GQDs@ZIF-8@NIP and N, S-GQDs@ZIF-8@MIP with OA. (**c**) UV-Vis absorption spectra of OA-a, N, S-GQDs@ZIF-8@MIP before-c and after-b interactions with OA; fluorescence excitation-d, and emission-e spectra of the N, S-GQDs@ZIF-8@MIP. (**d**) Fluorescence response of N, S-GQDs@ZIF-8@MIP (a), N, S-GQDs@MIP (b) and N, S-GQDs@ZIF-8@NIP (c) under different contact times.

Under the presence of octopamine, N, S-GQDs@ZIF-8@MIP revealed fluorescence enhancement. For this phenomenon, a fluorescence enhancement mechanism was proposed to explain the interaction between octopamine and N, S-GQDs@ZIF-8@MIP. First, as displayed in Figure 3c, the UV absorption spectrum of octopamine (a) did not overlap with the emission peak for N, S-GQDs@ZIF-8@MIP(e) of the fluorophore, which eliminated the possibility of fluorescence resonance energy transfer (FRET). Octopamine is a well-known aromatic compound with a large π bond benzene ring that can easily generate photoelectrons. Meanwhile, octopamine also has common electron-donating groups with amino and hydroxyl groups that can help the electron transfer from octopamine to fluorescent substances. Thus, the mechanism of fluorescence enhancement was speculated as a process that mostly relies on the electron transfer from octopamine to N, S-GQDs@ZIF-8@MIP. To further confirm this structure, the fluorescence lifetime of the materials was measured using time-resolved spectroscopy, which was used to monitor the emission of the 520 nm fluorophore in the absence and presence of the analyte. The results are provided in the supporting information file (Figure S2). According to the above analysis, the main mechanism for the fluorescence enhancement of N, S-GQDs@ZIF-8@MIP may be photo-induced electron transfer (PET).

To investigate the absorption ability of N, S-GQDs@ZIF-8@MIP to OA, experiments were designed using N, S-GQDs@ZIF-8@NIP and N, S-GQDs@MIP as comparisons.

As illustrated in Figure 3d, the absorption kinetics of N, S-GQDs@ZIF-8@MIP (a), N, S-GQDs@ZIF-8@NIP (b), and N, S-GQDs@MIP (c) were investigated by determining the variation in the fluorescence response (F/F_0) from 0 to 70 min. With a change in the adsorption time, the F/F_0 showed a remarkable increase. The adsorption of OA by N, S-GQDs@ZIF-8@MIP occurred in the first 30 min, and then slowly increased between 30 min and 40 min. After 40 min, there was no noticeable alteration. Similar absorption trends were observed for N, S-GQDs@ZIF-8@NIP. However, N, S-GQDs@ZIF-8@MIP displayed a higher F/F_0 value than N, S-GQDs@ZIF-8@NIP under the same conditions, which may be ascribed to the specific binding between OA and the -OH/-NH_2 groups of the imprinting recognition sites on the N, S-GQDs@ZIF-8@MIP. Meanwhile, the absorption performance of N, S-GQDs@ZIF-8@MIP was better than that of N, S-GQDs@MIP. Additionally, compared to N, S-GQDs@MIP, the adsorption equilibrium time of N, S-GQDs@ZIF-8@MIP was greatly shortened from 50 min to 40 min. These results indicated that the introduction of high surface area ZIF-8 can provide a platform for the MIP membrane, which can allow more imprinted sites on the surface and enhance the adsorption capacity of materials. The excellent OA adsorption by N, S-GQDs@ZIF-8@MIP illustrated that the introduction of high-surface-area ZIF-8 indeed improved the adsorption competence of the sensor.

3.4. Establishment of Fluorescence Sensing Detection for OA

To evaluate the dose–response curves of the prepared fluorescent materials for OA, N, S-GQDs@ZIF-8@MIP and N, S-GQDs@ZIF-8@NIP were dispersed in OA ethanol solution with various concentrations from 0.1 to 10 mg L^{-1}. The fluorescence intensity of the polymers gradually increased with an increase in the concentration of OA, and the results are shown in Figure 4. Compared to N, S-GQDs@ZIF-8@NIP, N, S-GQDs@ZIF-8@MIP produced a greater enhancement effect at the same OA concentration. The relationship between the fluorescence response of the sensor and the concentration of the OA conformed to the following equation [36]:

$$F/F_0 = 1 + K_{SV}\, C_{OA}$$

where F_0 and F are the fluorescence intensity in the absence and presence of OA, respectively; K_{SV} represents the constant; and C_{OA} is the concentration of OA. Herein, the imprinting factor IF ($K_{SV,\ N,\ S\text{-}GQDS@ZIF\text{-}8@MIP}/K_{SV,\ N,\ S\text{-}GQDS@ZIF\text{-}8@NIP}$) was used as an important indicator for evaluating the binding ability of the N, S-GQDs@ZIF-8@MIP sensor.

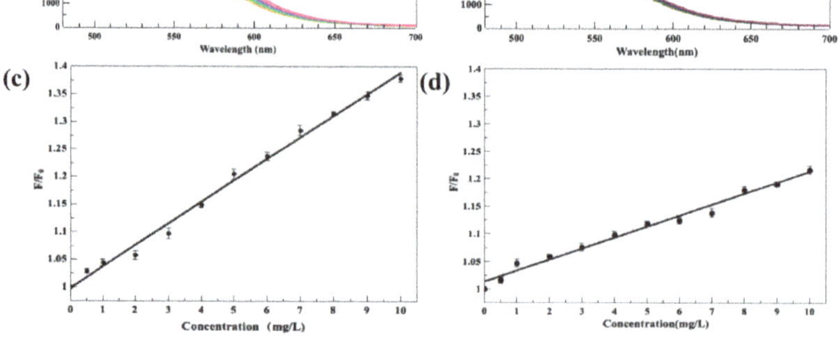

Figure 4. Fluorescence emission spectra of N, S-GQDs@ZIF-8@MIP (**a**) and N, S-GQDs@ZIF-8@NIP (**b**) under the different OA concentrations, the linear section of N, S-GQDs@ZIF-8@MIP (**c**) and N, S-GQDs@ZIF-8@NIP (**d**).

As expected, the fluorescence response sharply increased along with an increase in OA concentration (0.1–10 mg L^{-1}), which can be seen in Figure 4a. The corresponding equation with a good linear relationship of $F/F_0 = 0.039\ C_{OA} + 0.9984$ was obtained for N, S-GQDs@ZIF-8@MIP with correlation coefficients of 0.9933. The limit of detection (LOD) was determined to be 0.062 mg L^{-1}, and was calculated following the standard 3σ criteria (3σ/S) [37]. As shown in Figure 4b, the corresponding standard linear equation of N, S-GQDs@ZIF-8@NIP was $F/F_0 = 0.02\ C_{OA} + 1.013$ with correlation coefficients of 0.983. As displayed in Figure 4, under identical conditions, the N, S-GQDs@ZIF-8@MIP exhibited greater fluorescence enhancement than N, S-GQDs@ZIF-8@NIP due to the absence of OA-imprinted cavities in N, S-GQDs@ZIF-8@NIP. Under optimal conditions, the calculated *IF* was 1.95, which indicated that the constructed N, S-GQDs@ZIF-8@MIP had an excellent affinity for OA.

3.5. Selectivity of N, S-GQDs@ZIF-8@MIP

To investigate the selectivity of N, S-GQDs@ZIF-8@MIP(MIP) towards OA, a series of experiments was carried out using L-Try, L-Tyr, L-His, and DA as the structural analogues of OA. The structures of the analogues and OA are provided in Figure 5a. Additionally, the fluorescence responses (F/F_0) of MIP and N, S-GQDs@ZIF-8@NIP(NIP) to OA, L-Try, L-Tyr, and L-His are presented in Figure 5b. MIP clearly had higher adsorption capabilities for OA in the selectivity experiments and displayed little sensitivity to non-template molecules (L-Try, L-Tyr, L-His, and DA). The fluorescence response of NIP to OA was as weak as that of other structural analogues (L-Try, L-Tyr, L-His, and DA) because of the lack of specific imprinting sites.

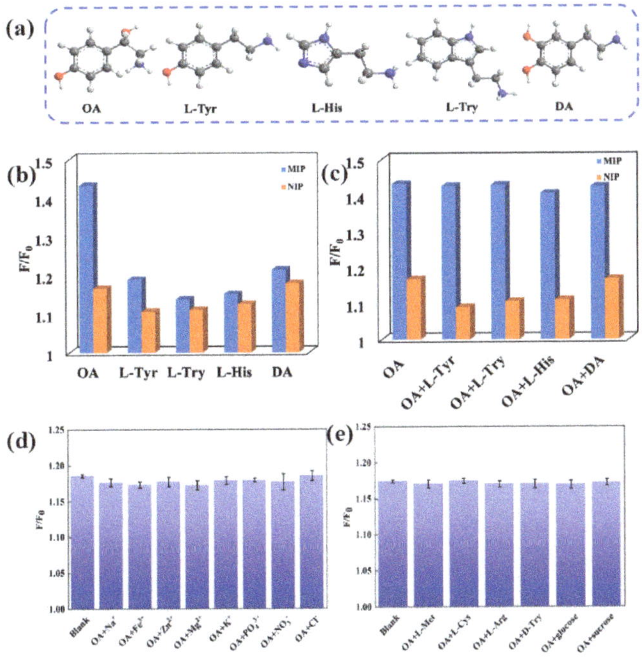

Figure 5. (**a**) The structure of OA, L-Try, L-Tyr, L-His, and DA. (**b**) Changes in the fluorescence intensity of N, S-GQDs@ZIF-8@MIP and N, S-GQDs@ZIF-8@NIP in response to OA and responses to the other analogs. (**c**) The effect of analogues on the binding of OA to the N, S-GQDs@ZIF-8@MIP and N, S-GQDs@ZIF-8@NIP. Influence of common coexisting ions (**d**) and other molecules (**e**) on the increasing fluorescence efficiency of N, S-GQDs@ZIF-8@MIP in the presence of OA.

Competitive adsorption experiments were also conducted by adding different structural analogues (L-Try, L-Tyr, L-His, and DA) into the solution containing OA. The fluorescence responses (F/F_0) of MIP and NIP were enhanced with almost the same efficiency (Figure 5c). The above results confirmed that MIP had excellent selectivity towards OA. On the basis of these observations, we can draw the following conclusion: MIP with imprinted cavities has a stronger affinity for OA.

Furthermore, some substances introduced by the sample pretreatments, including common metal ions (Na^+, K^+, Mg^{2+}, Fe^{3+}, and Zn^{2+}), amino acids (D-Try, L-Arg, L-Met, and L-Cys), and sugars (glucose and sucrose), may have affected the accuracy of the experimental results. Based on the presence of these interfering substances, an experiment was designed to confirm the specificity of the synthesized sensor in complex environments. We determined that even at concentrations up to 1 g L^{-1} (250-fold that of OA), these substances caused only a slight fluorescence change of MIP (Figure 5d,e).

These experiments confirmed that the obtained sensor has high sensitivity and excellent anti-interference capability for the recognition of OA. Thus, we conclude that the constructed MIP was a well-fluorescent material that can effectively detect OA in fermented samples.

3.6. OA Detection in Real Samples

To further evaluate the accuracy and applicability of the constructed N, S-GQDs@ZIF-8@MIP, a comparison between the constructed fluorescence-sensing method and classical detection method (HPLC-UV) was carried out. In this work, the actual samples were white vinegar and spirits. Since OA was not found in the two blank samples using the FL and HPLC methods, the recovery experiments were performed by spiking different concentrations of OA (1, 5, and 9 mg L^{-1}), and three parallel groups were set for the determination of each concentration of each sample. As outlined in Table 1, the relative recovery ratio and the relative standard deviation (RSD) of the sample solutions were determined under the condition that the same concentration was repeated three times, which was calculated from the following equation: relative recovery (%) = $(C_F - C_S/C_A) \times 100$, where C_F is the total OA found in a spiked sample, C_S is the OA in the original sample and C_A is the standard OA spiked in sample. In real sample detection, the FL method recoveries of OA were between 80.24% and 98.11%, and the RSD was in the range of 2.7–5.2. In addition, the HPLC method was used to verify the reliability of the developed FL test results, and the recoveries ranged from 80.38% to 87.17%, while RSD was in the range of 1.22–4.93. The results displayed that there was no significant difference (p value = 0.96 > 0.05) between this method and HPLC. By comparing the detection results under FL with those of HPLC, we determined that the N, S-GQDs@ZIF-8@MIP-based sensor can supply a well-pleasing platform for determination of OA in samples.

Table 1. The determination of OA-containing samples via the proposed N, S-GQDs@ZIF-8@MIP and HPLC-UV.

Samples	Added (mg L^{-1})	N, S-GQDs@ZIF-8@MIP			HPLC-UV		
		Found (mg L^{-1}) (n = 3)	Recovery% (Mean ± SD) (n = 3)	RSD% (n = 3)	Found (mg L^{-1}) (n = 3)	Recovery % (Mean ± SD) (n = 3)	RSD% (n = 3)
white vinegar	0.00	ND [a]	-	-	ND [a]	-	-
	1.00	0.87	86.86 ± 2.31	2.66	0.80	80.38 ± 3.96	4.93
	5.00	4.68	93.58 ± 1.64	1.75	4.30	85.90 ± 1.04	1.22
	9.00	8.00	88.92 ± 2.46	2.77	7.27	80.75 ± 2.60	3.23
wine	0.00	ND [a]	-	-	ND [a]	-	-
	1.00	0.91	91.08 ± 1.68	1.84	0.84	84.48 ± 2.24	2.65
	5.00	4.54	90.90 ± 4.13	4.54	4.36	87.17 ± 1.81	2.09
	9.00	8.29	92.07 ± 2.66	2.89	7.41	82.31 ± 2.36	2.87

[a] Not Detectable.

3.7. Method Performance Comparison

The performance of N, S-GQDs@ZIF-8@MIP-based fluorescence-sensing method was compared with the previous reported approaches for OA detection in Table 2. Apparently, the FL method constructed in this work displayed a lower detection limit, good precision (RSD) and linear range. The excellent performance of N, S-GQDs@ZIF-8@MIP-based sensing system benefited from the following aspects. Firstly, N, S-GQDs@ZIF-8 with extremely bright green light and large surface area was introduced as supporter, which can not only evidently improve the sensitivity of the sensor, but also reduced the aggregation quenching of N, S-GQDs. Secondly, the modification of the imprinting layer further enhanced the accuracy of the sensor in identifying target molecules, which was beneficial to avoiding tedious sample pre-processing and shortening analysis time. In summary, apart from the distinct advantages of specific recognition sites, the presented method was also simple in operation, rapid in response and cost-effective. Thence, the developed fluorescence imprinting sensor was more suitable and promising for the specific recognition and quantitative detection of trace OA in the fermented environment.

Table 2. Comparison of the proposed method with other methods reported in the literature.

Methods	Detector	LOD ($mg\ L^{-1}$)	Linear Range ($mg\ L^{-1}$)	Recovery (%)	RSD (%)	Advantages	Limitations	Reference
Electrochemical	octopamine tyramine	0.153 0.041	0.0153–6.12 0.013–0.034	98.5–104.7 102.2–103.1	5.9–6.1 5.6–6.4	fast sensitive	The scope of use is limited	[5]
ELISA	octopamine	0.59	2.259–15.3	96.2–106.4	-	high reproducibility	Low sensitivity and specificity	[6]
FSCV	octopamine	0.336	-	-	-	sensitive	Low accuracy, easily disturbed	[7]
HPLC	synephrine octopamine	0.033 0.092	0.2–1.2 0.2–1.2	98.2–101.5	1.5–2.5	wide range of applications	Costly, complex preprocessing	[8]
Fluorescence (UCNPs@ZIF-8@MIP)	octopamine	0.081	0.1–10	81.75–90.63	2.56–5.13	simple preprocessing	UCNP synthesis conditions were complex	[28]
Fluorescence (N,S-GQDs@ZIF-8@MIP)	octopamine	0.062	0.1–10	86.86–93.58	1.75–4.54	fast, sensitive, high detection limits		This work

4. Conclusions

In this paper, an efficient one-pot strategy was proposed to construct a novel N, S-GQDs@ZIF-8@MIP sensor that can achieve OA detection through photo-induced electron transfer. Through the introduction of N, S-GQDs with unique fluorescence properties, the obtained N, S-GQDs@ZIF-8@MIP presented a more significant fluorescence response to trace OA. The modification of molecular imprinting with the special recognition sites gave the sensor good selectivity. Meanwhile, the use of ZIF-8 with a high specific surface area improved the mass transfer rate of the N, S-GQDs@ZIF-8@MIP. The synthesized N, S-GQDs@ZIF-8@MIP sensor has the merits of favorable selectivity and higher sensitivity. More importantly, N, S-GQDs@ZIF-8@MIP was successfully applied to the detection of OA in real samples. Therefore, N, S-GQDs@ZIF-8@MIP has a promising prospect for OA detection in fermented foods.

Supplementary Materials: The following are available online at https://www.mdpi.com/article/10.3390/foods11091348/s1, Figure S1: The high-angle annular dark-field (HAADF) image of N, S-GQDs@ZIF-8, and corresponding element mappings of C (B), N (C), O (D), S (E), and Zn (F); Figure S2: Fluorescence emission decay curves of N, S-GQDs@ZIF-8@MIP alone and N, S-GQDs@ZIF-8@MIP with 40 mg mL^{-1} concentrations of OA under excitation at 460 nm. The concentration of the N, S-GQDs@ZIF-8@MIP was 2 mg mL^{-1}, Figure S3: The stability of N, S-GQDs@ZIF-8@MIP from 0 to 720 min at room temperature; Table S1: Optimization of addition ratio; Table S2: Optimization of ZIF-8; Table S3: Optimization of N, S-GQDs dosage.

Author Contributions: Y.G., conceptualization, data curation, validation, visualization, writing—original draft, and writing—review and editing; G.Y., data curation and writing—reviewing; X.H., investigation and validation; J.Z., investigation and validation; G.F., project administration and supervision. All authors have read and agreed to the published version of the manuscript.

Funding: This work was supported by the National Key Research and Development Program of China (project No. 2017YFC1600803).

Data Availability Statement: The datasets generated for this study are available on request from the corresponding author.

Conflicts of Interest: The authors declare that they have no competing interests.

References

1. Tofalo, R.; Perpetuini, G.; Schirone, M.; Suzzi, G. Biogenic Amines: Toxicology and Health Effect. *Encycl. Food Health* **2015**, *71*, 424–429.
2. Elias, D.O.; Land, B.R.; Mason, A.C.; Hoy, R.R. Measuring and quantifying dynamic visual signals in jumping spiders. *J. Comp. Physiol. A Sens. Neural Behav. Physiol.* **2006**, *192*, 785–797. [CrossRef] [PubMed]
3. Wójcik, W.; Łukasiewicz, M.; Puppel, K. Biogenic amines: Formation, action and toxicity—A review. *J. Sci. Food Agric.* **2020**, *101*, 2634–2640. [CrossRef] [PubMed]
4. Thevis, M.; Koch, A.; Sigmund, G.; Thomas, A.; Schänzer, W. Analysis of octopamine in human doping control samples. *Biomed. Chromatogr.* **2011**, *26*, 610–615. [CrossRef] [PubMed]
5. Zhang, Y.; Zhang, M.; Wei, Q.; Gao, Y.; Guo, L.; Al-Ghanim, K.A.; Mahboob, S.; Zhang, X. An Easily Fabricated Electrochemical Sensor Based on a Graphene-Modified Glassy Carbon Electrode for Determination of Octopamine and Tyramine. *Sensors* **2016**, *16*, 535. [CrossRef] [PubMed]
6. Moczko, E.; Díaz, R.; Rivas, B.; García, C.; Pereira, E.; Piletsky, S.; Cáceres, C. Molecularly Imprinted Nanoparticles Assay (MINA) in Pseudo ELISA: An Alternative to Detect and Quantify Octopamine in Water and Human Urine Samples. *Polymers* **2019**, *11*, 1497. [CrossRef]
7. Pyakurel, P.; Champaloux, E.P.; Venton, B.J. Fast-Scan Cyclic Voltammetry (FSCV) Detection of Endogenous Octopamine in Drosophila melanogaster Ventral Nerve Cord. *ACS Chem. Neurosci.* **2016**, *7*, 1112–1119. [CrossRef]
8. Pawar, R.S.; Sagi, S.; Leontyev, D. Analysis of bitter orange dietary supplements for natural and synthetic phenethylamines by LC–MS/MS. *Drug Test. Anal.* **2020**, *12*, 1241–1251. [CrossRef]
9. Kang, J.H.; Kim, C. Colorimetric detection of iron and fluorescence detection of zinc and cadmium by a chemosensor containing a bio-friendly octopamine. *Photochem. Photobiol. Sci.* **2018**, *17*, 442–452. [CrossRef]
10. Gupta, V.; Chaudhary, N.; Srivastava, R.; Sharma, G.D.; Bhardwaj, R.; Chand, S. Luminscent Graphene Quantum Dots for Organic Photovoltaic Devices. *J. Am. Chem. Soc.* **2011**, *133*, 9960–9963. [CrossRef]
11. Du, Y.; Guo, S. Chemically doped fluorescent carbon and graphene quantum dots for bioimaging, sensor, catalytic and photoelectronic applications. *Nanoscale* **2015**, *8*, 2532–2543. [CrossRef] [PubMed]
12. Kundu, S.; Yadav, R.M.; Narayanan, T.N.; Shelke, M.V.; Vajtai, R.; Ajayan, P.M.; Pillai, V.K. Synthesis of N, F and S Co-Doped Graphene Quantum Dots. *Nanoscale* **2015**, *7*, 11515. [CrossRef] [PubMed]
13. Qu, D.; Zheng, M.; Zhang, L.; Zhao, H.; Xie, Z.; Jing, X.; Haddad, R.E.; Fan, H.; Sun, Z. Formation mechanism and optimization of highly luminescent N-doped graphene quantum dots. *Sci. Rep.* **2014**, *4*, 5294. [CrossRef] [PubMed]
14. Li, X.; Lau, S.P.; Tang, L.; Ji, R.; Yang, P. Sulphur doping: A facile approach to tune the electronic structure and optical properties of graphene quantum dots. *Nanoscale* **2014**, *6*, 5323–5328. [CrossRef]
15. Li, S.; Li, Y.; Cao, J.; Zhu, J.; Fan, L.; Li, X. Sulfur-Doped Graphene Quantum Dots as a Novel Fluorescent Probe for Highly Selective and Sensitive Detection of Fe^{3+}. *Anal. Chem.* **2014**, *86*, 10201–10207. [CrossRef]
16. Xia, C.; Hai, X.; Chen, X.-W.; Wang, J.-H. Simultaneously fabrication of free and solidified N, S-doped graphene quantum dots via a facile solvent-free synthesis route for fluorescent detection. *Talanta* **2017**, *168*, 269–278. [CrossRef]
17. Yaghi, O.M.; Li, G.; Li, H. Selective binding and removal of guests in a microporous metal–organic framework. *Nature* **1995**, *378*, 703–706. [CrossRef]
18. Pham, T.; Forrest, K.A.; Franz, D.M.; Space, B. Experimental and theoretical investigations of the gas adsorption sites in rht-metal–organic frameworks. *CrystEngComm* **2017**, *19*, 4646–4665. [CrossRef]
19. Li, J.; Wang, X.; Zhao, G.; Chen, C.; Chai, Z.; Alsaedi, A.; Hayat, T.; Wang, X. Metal–organic framework-based materials: Superior adsorbents for the capture of toxic and radioactive metal ions. *Chem. Soc. Rev.* **2018**, *47*, 2322–2356. [CrossRef]
20. Altintas, C.; Avci, G.; Daglar, H.; Azar, A.N.V.; Velioglu, S.; Erucar, I.; Keskin, S. Database for CO_2 Separation Performances of MOFs Based on Computational Materials Screening. *ACS Appl. Mater. Interfaces* **2018**, *10*, 17257–17268. [CrossRef]
21. Zhou, Y.; Li, X.; Pan, Z.; Ye, B.; Xu, M. Determination of Malachite Green in Fish by a Modified MOF-Based Electrochemical Sensor. *Food Anal. Methods* **2019**, *12*, 1246–1254. [CrossRef]
22. Zhu, Q.-L.; Xu, Q. Metal-organic framework composites. *Chem. Soc. Rev.* **2014**, *43*, 5468–5512. [CrossRef]
23. Tripathy, S.P.; Subudhi, S.; Parida, K. Inter-MOF hybrid (IMOFH): A concise analysis on emerging core–shell based hierarchical and multifunctional nanoporous materials. *Coord. Chem. Rev.* **2021**, *434*, 213786. [CrossRef]

24. Xu, L.; Fang, G.; Liu, J.; Pan, M.; Wang, R.; Wang, S. One-pot synthesis of nanoscale carbon dots-embedded metal–organic frameworks at room temperature for enhanced chemical sensing. *J. Mater. Chem. A* **2016**, *4*, 15880–15887. [CrossRef]
25. Chen, L.; Wang, X.; Lu, W.; Wu, X.; Li, J. Molecular imprinting: Perspectives and applications. *Chem. Soc. Rev.* **2016**, *45*, 2137–2211. [CrossRef]
26. Malik, M.I.; Shaikh, H.; Mustafa, G.; Bhanger, M.I. Recent Applications of Molecularly Imprinted Polymers in Analytical Chemistry. *Sep. Purif. Rev.* **2018**, *48*, 179–219. [CrossRef]
27. Rutkowska, M.; Płotka-Wasylka, J.; Morrison, C.; Wieczorek, P.P.; Namiesnik, J.; Marć, M. Application of molecularly imprinted polymers in analytical chiral separations and analysis. *TrAC Trends Anal. Chem.* **2018**, *102*, 91–102. [CrossRef]
28. Caoa, Y.; Hua, X.; Zhaoa, T.; Maoa, Y.; Fanga, G.; Wangab, S. A core-shell molecularly imprinted optical sensor based on the upconversion nanoparticles decorated with Zinc-based metal-organic framework for selective and rapid detection of octopamine. *Sensors Actuators B Chem.* **2020**, *326*, 128838. [CrossRef]
29. Hu, X.; Cao, Y.; Tian, Y.; Qi, Y.; Fang, G.; Wang, S. A molecularly imprinted fluorescence nanosensor based on upconversion metal–organic frameworks for alpha-cypermethrin specific recognition. *Mikrochim. Acta* **2020**, *187*, 1–10. [CrossRef]
30. Qu, D.; Sun, Z.; Zheng, M.; Li, J.; Zhang, Y.; Zhang, G.; Zhao, H.; Liu, X.; Xie, Z. Three Colors Emission from S.N. Co-doped Graphene Quantum Dots for Visible Light H_2Production and Bioimaging. *Adv. Opt. Mater.* **2015**, *3*, 360–367. [CrossRef]
31. Chen, S.; Hai, X.; Xia, C.; Chen, X.-W.; Wang, J.-H. Preparation of Excitation-Independent Photoluminescent Graphene Quantum Dots with Visible-Light Excitation/Emission for Cell Imaging. *Chem. A Eur. J.* **2013**, *19*, 15918–15923. [CrossRef]
32. Qu, D.; Zheng, M.; Du, P.; Zhou, Y.; Zhang, L.; Li, D.; Tan, H.; Zhao, Z.; Xie, Z.; Sun, Z. Highly luminescent S, N co-doped graphene quantum dots with broad visible absorption bands for visible light photocatalysts. *Nanoscale* **2013**, *5*, 12272–12277. [CrossRef]
33. Xu, H.; Zhou, S.; Xiao, L.; Yuan, Q.; Gan, W. Time-efficient syntheses of nitrogen and sulfur co-doped graphene quantum dots with tunable luminescence and their sensing applications. *RSC Adv.* **2016**, *6*, 36554–36560. [CrossRef]
34. Cravillon, J.; Münzer, S.; Lohmeier, S.-J.; Feldhoff, A.; Huber, K.; Wiebcke, M. Rapid Room-Temperature Synthesis and Characterization of Nanocrystals of a Prototypical Zeolitic Imidazolate Framework. *Chem. Mater.* **2009**, *21*, 1410–1412. [CrossRef]
35. He, L.; Wang, T.; An, J.; Li, X.; Zhang, L.; Li, L.; Li, G.; Wu, X.; Su, Z.; Wang, C. Carbon nanodots@zeolitic imidazolate framework-8 nanoparticles for simultaneous pH-responsive drug delivery and fluorescence imaging. *CrystEngComm* **2014**, *16*, 3259–3263. [CrossRef]
36. Yang, J.; Feng, W.; Liang, K.; Chen, C.; Cai, C. A novel fluorescence molecularly imprinted sensor for Japanese encephalitis virus detection based on metal organic frameworks and passivation-enhanced selectivity. *Talanta* **2020**, *212*, 120744. [CrossRef]
37. Analytical Methods Committee Recommendations for the definition, estimation and use of the detection limit. *Analyst* **1987**, *112*, 199–204. [CrossRef]

foods

MDPI

Article

A Turn-Off Fluorescent Biomimetic Sensor Based on A Molecularly Imprinted Polymer-Coated Amino-Functionalized Zirconium (IV) Metal–Organic Framework for the Ultrasensitive and Selective Detection of Trace Oxytetracycline in Milk

Xiaohui Wang [1], Chang Liu [2], Yichuan Cao [1], Lin Cai [1], Haiyang Wang [1] and Guozhen Fang [1,*]

1 State Key Laboratory of Food Nutrition and Safety, Tianjin University of Science and Technology, Tianjin 300457, China; xhw2022tjin@163.com (X.W.)

2 School of Food Science, Henan Institute of Science and Technology, Xinxiang 453003, China

* Correspondence: fangguozhen@tust.edu.cn; Tel.: +86-22-60912493

Abstract: Developing sensitive and effective methods to monitor oxytetracycline residues in food is of great significance for maintaining public health. Herein, a fluorescent sensor (NH_2-UIO-66 (Zr)@MIP) based on a molecularly imprinted polymer-coated amino-functionalized zirconium (IV) metal–organic framework was successfully constructed and first used for the ultrasensitive determination of oxytetracycline. NH_2-UIO-66 (Zr), with a maximum emission wavelength of 455 nm under 350 nm excitation, was prepared using a microwave-assisted heating method. The NH_2-UIO-66 (Zr)@MIP sensor with specific recognition sites for oxytetracycline was then acquired by modifying a molecularly imprinted polymer on the surface of NH_2-UIO-66 (Zr). The introduction of NH_2-UIO-66 (Zr) as both a signal tag and supporter can strengthen the sensitivity of the fluorescence sensor. Thanks to the combination of the unique characteristics of the molecularly imprinted polymer and NH_2-UIO-66 (Zr), the prepared sensor not only exhibited a sensitive fluorescence response, specific identification capabilities and a high selectivity for oxytetracycline, but also showed good fluorescence stability, satisfactory precision and reproducibility. The fabricated sensor displayed a fluorescent linear quenching in the OTC concentration range of 0.05–40 μg mL^{-1}, with a detection limit of 0.012 μg mL^{-1}. More importantly, the fluorescence sensor was finally applied for the detection of oxytetracycline in milk, and the results were comparable to those obtained using the HPLC approach. Hence, the NH_2-UIO-66 (Zr)@MIP sensor possesses great application potential for the accurate evaluation of trace oxytetracycline in dairy products.

Keywords: turn-off fluorescent biomimetic sensor; amino-functionalized zirconium (IV) metal–organic framework; molecularly imprinted polymer; oxytetracycline

Citation: Wang, X.; Liu, C.; Cao, Y.; Cai, L.; Wang, H.; Fang, G. A Turn-Off Fluorescent Biomimetic Sensor Based on A Molecularly Imprinted Polymer-Coated Amino-Functionalized Zirconium (IV) Metal–Organic Framework for the Ultrasensitive and Selective Detection of Trace Oxytetracycline in Milk. *Foods* **2023**, *12*, 2255. https://doi.org/10.3390/foods12112255

Academic Editor: Thierry Noguer

Received: 18 May 2023
Revised: 28 May 2023
Accepted: 1 June 2023
Published: 3 June 2023

1. Introduction

Luminescent metal organic frameworks (LMOFs), as rapidly developing novel-type organic–inorganic porous crystal materials, have displayed wide application potential in sensing fields owing to their high specific surface area and porosity, intrinsic luminescence properties, as well as structure multiplicity [1,2]. Among them, NH_2-UIO-66 (Zr) is a kind of stable zirconium-based metal–organic framework (Zr-MOF), and has been developed to construct fluorescent sensors in recent years on account of its excellent fluorescence (FL) characteristics, chemical stability and large surface area [3–6]. For instance, Zhu et al. fabricated a NH_2-UIO-66 (Zr)-based sensor for the sensitive FL detection of the fluoride anion in water medium [6]. NH_2-UIO-66 (Zr) was successfully utilized as a fluorescent probe for sensitively detecting the phosphate anion in aqueous media via the FL enhancement effect [7]. Wang et al. designed a novel ratio FL probe based on fluorescent Zr-MOF for the highly sensitive determination of dopamine and reduced glutathione [5]. Wang et al. stated clearly that the NH_2-UIO-66 fluorescent sensor was triumphantly prepared and applied for the FL sensing of tetracyclines in milk [8]. Although these NH_2-UIO-66 (Zr)-based optical sensors showed gratifying FL sensing performances, there were still some limitations, such as a relatively low selectivity for target analytes and a weak anti-interference capability in elaborate actual samples. As a result, the establishment of LMOF-based sensors with specific recognition abilities and high selectivity is a vital method by which to achieve the accurate evaluation of target analytes in real samples.

Molecular imprinting technology (MIT) is a promising molecular simulation recognition technique that mimics the specific antigen–antibody or enzyme–substrate binding processes, and has attracted great attention in the past few decades. The formed molecularly imprinted polymer (MIP) not only possesses the ability to specifically recognize the detected molecules, but also has a high adsorption efficiency and strong stability, which makes it widely used in various fields (such as catalysis, separation, drug delivery, gas sensing, chemical and biological sensing) [9–12]. In addition, MIP has been applied in optical sensors to enhance the selectivity of sensors. To date, some research has been conducting regarding the combination of NH_2-UIO-66 (Zr) with MIP in order to design fluorescent sensors, in which the introduction of MIP can promote the generation of customized binding sites complementary to the target analytes in terms of size, shape and functional group, thus improving the selectivity of the obtained sensor. For example, a MOF@MIP FL sensor (NH_2-UIO-66@MIP), synthetized by encapsulating NH_2-UIO-66 into the MIP, was successfully exploited for the ultrasensitive and highly selective detection of melatonin in juice [13]. A LMOF-based sensor (NH_2-UIO-66/MIP) was constructed to selectively and effectively detect trace levels of 4-nitrophenol in drinkable water, as well as in environmental water [14]. A turn-on FL sensor (MIP/NH_2-UIO-66) was made via decorating MIP on Zr-LMOF and was employed for the FL sensing of chloramphenicol in animal-derived food [15]. Therefore, the integration of LMOFs and MIP in order to construct novel fluorescent sensors can supply an efficacious means by which to perform food safety supervision, such as monitoring the residual trace antibiotics.

For many decades, the abuse of antibiotics has been regarded as a global concern because antibiotic residues may constitute a threat to human health [16,17]. Oxytetracycline (OTC), a tetracycline antibiotic, has been extensively utilized in the remedy of infectious diseases induced by pathogenic microorganisms due to its broad-spectrum antimicrobial activity, effectiveness and low cost [18]. Nevertheless, the overuse of OTC will lead to its accumulation in the environment and foodstuffs, thus causing a series of potential hazards, such as anaphylactic reactions, kidney and liver damage, decreased human immune function, as well as the emergence of antibiotic-resistant bacteria [19–21]. To guarantee consumer safety, the European Union (EU) and China have dictated the maximum OTC residue limits (MRL) in milk products, which are 225 nM and 100 $\mu g\ kg^{-1}$, respectively [22–24]. At present, plentiful analytical techniques, including capillary electrophoresis (CE) [25], liquid chromatography–tandem mass spectrometry (LC–MS/MS) [26], enzyme-linked immunosorbent assay (ELISA) [27], high-performance liquid chromatography (HPLC) [28], fluorometric assay [29] and so on, have been employed to quantitatively determine OTC. In these methods, FL sensing approaches have shown considerable superiorities over traditional analysis methods, such as a convenience of operation, an inexpensive cost, high sensibility and a rapid response, which are more appropriate for the FL sensing detection of OTC [30–32]. Hence, taking account of the current status of OTC, it is still indispensable to construct a new, reliable and efficient FL sensing method with which to monitor OTC in food. In view of the above research and analysis, the FL sensing strategy based on the incorporation of LMOFs and MIP has potential regarding its ability to offer a sensitive and efficacious technique with which to determine OTC.

Although some FL sensors that integrate the merits of LMOFs and MIP have been designed [13–15,33–35], hitherto, there has been no report on a fluorescent sensor based on MIP-coated NH_2-UIO-66 (Zr) for the selective and ultrasensitive FL monitoring of OTC.

Consequently, in this work, a turn-off FL sensor (expressed as NH_2-UIO-66 (Zr)@MIP) that combines a large surface area and the unique fluorescence properties of NH_2-UIO-66 (Zr), as well as the high specificity of the imprinted layer, was developed for an assay of OTC using the inner filter effect (IFE). Herein, the NH_2-UIO-66 (Zr) with an intense blue FL emission, as a signal tag and substrate material, was prepared via a simple microwave-assisted heating process and then covered with a molecularly imprinted layer. After removing the OTC template molecule, the NH_2-UIO-66 (Zr)@MIP sensor with custom-made binding sites for the OTC was obtained. The modification of the MIP granted the sensor high selectivity, adsorption affinity and binding efficiency to OTC. Meanwhile, NH_2-UIO-66 (Zr), as a luminous center and a carrier, endowed the proposed sensor with a desirable FL signal output and response sensitivity. Furthermore, its relatively uniform morphology, good optical properties, superior FL sensitivity, outstanding adsorption and binding performance, as well as its excellent specificity, were demonstrated via different characterizations. The established sensor was successfully used in the highly sensitive and selective determination of OTC in milk samples, confirming its practical applicability.

2. Materials and Methods

2.1. Materials and Apparatuses

2-aminoterephthalic acid (ATA, 98%), tetracycline (TET, ≥98%), doxycycline (DOX, ≥98%), oxytetracycline (OTC, ≥98%), chlortetracycline (CTC, ≥98%), chloramphenicol (CAP, 98%), tyrosine (Tyr), histidine (His), serine (Ser), phenylalanine (Phe), arginine (Arg) and cysteine (Cys) were provided by Shanghai Macklin Biochemical Co., Ltd. (Shanghai, China). Zirconium chloride ($ZrCl_4$, 99%) was purchased from Beijing Yinuokai Technology Co., Ltd. (Beijing, China). Methacrylic acid (MAA, 99%), 2,2-azobisisobutyronitrile (AIBN, 99%) and ethylene glycol dimethacrylate (EGDMA, 98%) were obtained from TCI Development Co., Ltd. (Shanghai, China). N, N-dimethylformamide (DMF), ethanol, glucose, sucrose, $MnCl_2 \cdot 4H_2O$, $MgCl_2 \cdot 6H_2O$, $Zn(NO_3)_2 \cdot 6H_2O$, $CaCl_2$, NaCl and KCl were supplied by Sinopharm Chemical Reagent Co., Ltd. (Shanghai, China). All reagents and chemicals used in this work were of at least analytical grade. The apparatuses utilized in this study are described in detail in the supporting information.

2.2. The Synthesis of NH_2-UIO-66 (Zr)

The preparation process of NH_2-UIO-66 (Zr) refers to the microwave-assisted heating assay employed in a previous study with minor changes [6]. Briefly, 669 mg of $ZrCl_4$ and 500 mg of ATA were firstly dissolved in 60 mL of DMF, followed by ultrasonic treatment for 1 h. Next, the aforementioned reaction precursor solution was placed in a stainless steel Teflon-lined autoclave and heated to 120 °C for 24 h. After natural cooling, the deposition was attained via centrifugation, distributed in ethanol (30 mL), and heated at 100 °C for 30 min in a microwave synthesis reaction device. Subsequently, after dropping to ambient temperature, the obtained solid was centrifuged, and then the above microwave heating activation process was carried out again. Ultimately, the yellow product was acquired via vacuum drying (60 °C for 6 h) for later tests.

2.3. The Fabrication of NH_2-UIO-66 (Zr) Imprinted Polymers

NH_2-UIO-66 (Zr)@MIP was fabricated according to a previous study, with slight modifications [15]. In the typical procedure, NH_2-UIO-66 (Zr) (fluorescent probe, 50 mg) and OTC (template, 0.2 mmol) were completely dispersed in 30 mL of ethanol using ultrasonic treatment. Then, after the addition of MAA (monomer, 1.2 mmol), the mixture was pre-polymerized for 50 min under continuous stirring. Subsequently, EGDMA (cross-linking agent, 2 mmol) and AIBN (initiator, 20 mg) were successively introduced into the above synthesis system and stirred for 15 h in a water bath at 65 °C. Finally, the acquired polymer was centrifuged, eluted repeatedly via water/ethanol (v/v, 1/9) to expunge OTC and dried under vacuum at 50 °C for 7 h. Meanwhile, the non-imprinting polymer (NH_2-

UIO-66 (Zr)@NIP), as a contrast material, was prepared according to the above steps in addition to not adding OTC.

2.4. FL Sensing Procedure

In brief, 1 mg of NH_2-UIO-66 (Zr)@MIP or NH_2-UIO-66 (Zr)@NIP was scattered in 4 mL of PBS buffer solution (pH = 7) with various concentrations of OTC. After incubation for 30 min, the FL determination was performed under an excitation wavelength of 350 nm and the FL intensity of the sensing system at 455 nm was obtained. The specific recognition performance was appraised by adding other antibiotics with a similar structure or possible coexisting interfering substances instead of OTC under identical experimental conditions. The measurements of each experiment were carried out in triplicate.

2.5. Real Samples Analysis

Fresh pure milk bought from a local supermarket (labeled as Milk 1, Milk 2, and Milk 3, according to brand difference) was used to assess the practicability of the proposed FL method based on NH_2-UIO-66 (Zr)@MIP, by spiking the proper concentration of OTC. Three parallel experiments were conducted on the milk samples of each concentration.

The spiked milk samples containing specific OTC concentrations were first prepared and placed overnight. Afterwards, the preprocessing of the samples was executed according to a previous study [36]. In brief, 1 mL milk samples with various OTC concentrations ($0\ \mu g\ mL^{-1}$, $1\ \mu g\ mL^{-1}$, $15\ \mu g\ mL^{-1}$ and $25\ \mu g\ mL^{-1}$) were vortically mingled with 4 mL of acetonitrile for 5 min. Subsequently, the mixture was centrifuged and the resulting supernatant was filtrated using a 0.22 μm microfilter to expunge protein precipitates. Finally, the above samples were dried using nitrogen and re-dispersed in 1 mL of PBS buffer solution (pH = 7) for FL determination.

2.6. HPLC Analysis

With regard to HPLC determination, the extraction, purification and analysis procedures of milk samples were carried out according to Chinese National Standards (GB/T 22990-2008) [37], which were provided in the supporting information.

3. Results and Discussion

3.1. Synthesis of NH_2-UIO-66 (Zr) and NH_2-UIO-66 (Zr)@MIP

As portrayed in Scheme 1, NH_2-UIO-66 (Zr) with a bright blue fluorescence and large surface area was first fabricated via a self-assembly strategy using Zr^{4+} and ATA as precursors. Subsequently, a new biomimetic FL sensor based on MIP-coated NH_2-UIO-66 (Zr) was prepared using the one-step aggregation method. The polymerization reaction was thermally initiated and completed in a water bath in the presence of OTC (template), NH_2-UIO-66 (Zr) (fluorophore and supporter), MAA (functional monomer), EGDMA (cross-linker) and AIBN (initiator). During this process, OTC molecules and functional monomers were combined by hydrogen bonds. After eliminating the imprinted template OTC by using a suitable eluent, the NH_2-UIO-66 (Zr)@MIP sensor with specific recognition cavities fitting the shape and functional group of OTC was finally achieved in order to selectively monitor the OTC. The existence of NH_2-UIO-66 (Zr) as both a signal transducer and carrier granted the proposed sensor outstanding FL characteristics and good response sensitivity. The introduction of a molecularly imprinted layer not only preserved the sensitive FL signal of NH_2-UIO-66 (Zr), but also endowed the obtained sensor with the ability to specifically recognize OTC. Consequently, the constructed NH_2-UIO-66 (Zr)@MIP sensor integrated these superiorities of LMOFs and MIP. Moreover, in order to obtain NH_2-UIO-66 (Zr)@MIP with an excellent fluorescent response and the ability to bind well to the OTC, the synthesis conditions, including the additive level of fluorophore, the molar ratio of OTC, MAA and EGDMA, and the polymerization time, were optimized (Tables S1–S3).

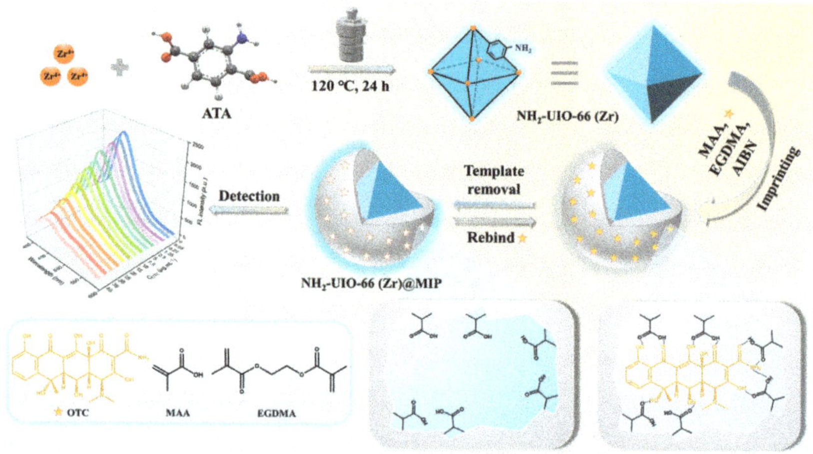

Scheme 1. Schematic diagram of the preparation of NH_2-UIO-66 (Zr) and NH_2-UIO-66 (Zr)@MIP.

3.2. Characterization

The surface morphologies and structures of the fabricated NH_2-UIO-66 (Zr), NH_2-UIO-66 (Zr)@MIP and NH_2-UIO-66 (Zr)@NIP were observed via SEM and TEM. SEM (Figure 1a), TEM (Figure 1d) and diameter distribution images (Figure S1) revealed that NH_2-UIO-66 (Zr) presented a uniform and smooth octahedron-like appearance, with a mean size of nearly 50 nm [38,39]. The elemental composition of NH_2-UIO-66 (Zr) was also explored via the energy-dispersive spectrometer (EDS) spectrum (Figure S2), confirming the coexistence of C, N, O and Zr elements in NH_2-UIO-66 (Zr). These results demonstrated the successful synthesis of NH_2-UIO-66 (Zr). As displayed in Figure 1b,c, the obtained polymers (NH_2-UIO-66 (Zr)@MIP and NH_2-UIO-66 (Zr)@NIP) appeared nearly spherical, with a particle size of approximately 250 nm (Figure S3), and their surface was rougher than that of NH_2-UIO-66 (Zr) due to the introduction of the molecular imprinting layer. The TEM images (Figure 1e,f) showed that the polymers basically maintained the frame of NH_2-UIO-66 (Zr) and possessed an archetypal core-shell structure, which evidenced that the polymers were successfully formed. In addition, the element mapping image of NH_2-UIO-66 (Zr)@MIP (Figure S4) showed the homogeneous distribution of the main characteristic elements in the polymer, preliminarily verifying the triumphant construction of the NH_2-UIO-66 (Zr)@MIP sensor.

The FT–IR spectra were recorded in order to analyze the surface chemistry of the synthetic materials. As presented in Figure 1g, in the spectrum of NH_2-UIO-66 (Zr), the characteristic peaks at 3492 and 3373 cm^{-1} can be attributed to the antisymmetric and symmetric stretching modes of -NH_2, respectively [8,40]. The absorption bands in the range of 1400–1500 cm^{-1} originated from a C=C stretching vibration in the benzene ring of the ATA ligand [14,15]. The stretching absorption peaks of O–Zr–O (760 and 663 cm^{-1}), Zr–O (483 cm^{-1}) and -OCO (1430 and 1384 cm^{-1}) were also observed, indicating that the dehydroxylation phase in NH_2-UIO-66 (Zr) was favorably formed thanks to the coordination of Zr^{4+} with ATA [6,41]. For NH_2-UIO-66 (Zr)@NIP, the absorption peaks at 3544 cm^{-1} (-OH), 2988 and 2953 cm^{-1} (C=H), 1731 cm^{-1} (C=O), 1455 cm^{-1} (C=C), and 1260 and 1159 cm^{-1} (C–O–C) symbolized the presence of MAA (functional monomer) and EGDMA (cross-linker). Meanwhile, the Zr–O vibration peaks (760, 663 and 483 cm^{-1}) corresponding to the NH_2-UIO-66 (Zr) appeared in NH_2-UIO-66 (Zr)@NIP, which demonstrated the successful preparation of the non-imprinted polymer doped with NH_2-UIO-66 (Zr) [13,14,42]. Additionally, the position and intensity of the characteristic peaks in the NH_2-UIO-66 (Zr)@MIP spectrum were very similar to those in the NH_2-UIO-66 (Zr)@NIP spectrum, implying that NH_2-UIO-66 (Zr)@MIP was triumphantly developed.

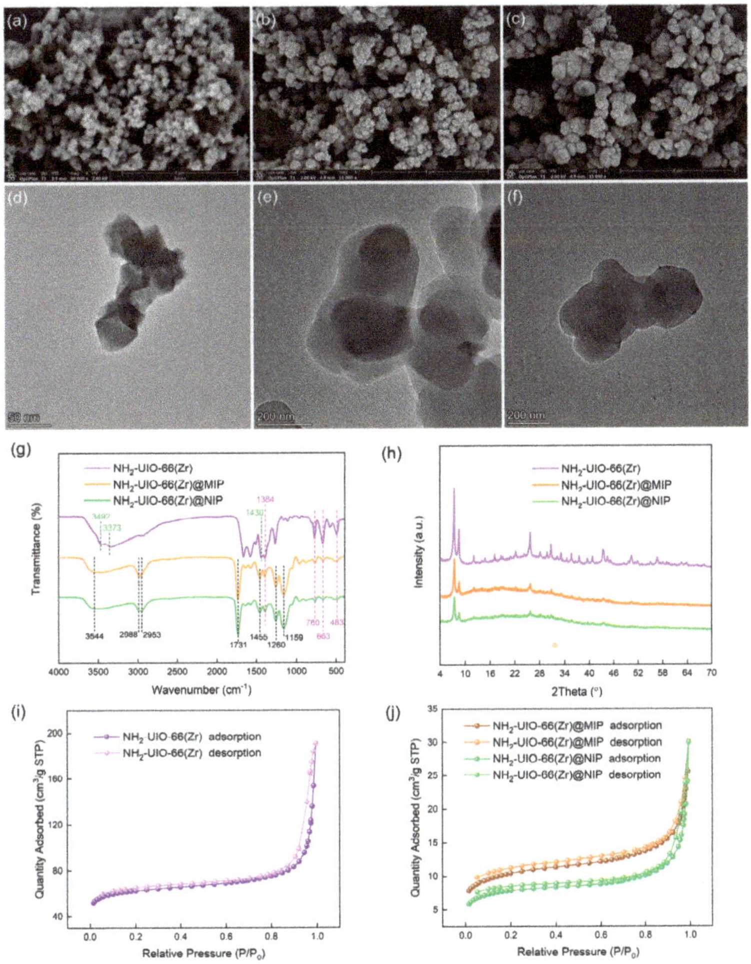

Figure 1. SEM images of (**a**) NH_2-UIO-66 (Zr), (**b**) NH_2-UIO-66 (Zr)@MIP and (**c**) NH_2-UIO-66 (Zr)@NIP. TEM images of (**d**) NH_2-UIO-66 (Zr), (**e**) NH_2-UIO-66 (Zr)@MIP and (**f**) NH_2-UIO-66 (Zr)@NIP. (**g**) FT–IR spectra and (**h**) XRD patterns of NH_2-UIO-66 (Zr), NH_2-UIO-66 (Zr)@MIP and NH_2-UIO-66 (Zr)@NIP. (**i**) Nitrogen adsorption and desorption curves of NH_2-UIO-66 (Zr). (**j**) Nitrogen adsorption and desorption curves of NH_2-UIO-66 (Zr)@MIP and NH_2-UIO-66 (Zr)@NIP.

To analyze the crystallographic structure of the synthesized materials, XRD data were collected and are depicted in Figure 1h. The characteristic diffraction peaks (2θ = 7.37°, 8.6°, 12.05°, 17.06°, 22.17° and 25.72°) of NH_2-UIO-66 (Zr) highly coincided with those reported in the literature, which showed that NH_2-UIO-66 (Zr) with good crystallinity was triumphantly fabricated [13,14,43]. Notably, in the XRD patterns of NH_2-UIO-66 (Zr)@MIP and NH_2-UIO-66 (Zr)@NIP, the characteristic absorption peaks associated with NH_2-UIO-66 (Zr) became unapparent, owing to the decoration of the surface molecularly imprinted shell, which helped to prove the successful preparation of the polymers [43,44].

In order to further investigate the surface characteristics of the resulting materials, the N_2 adsorption/desorption isotherms of NH_2-UIO-66 (Zr) and its imprinted polymer were evaluated, as shown in Figure 1i,j. The corresponding specific surface areas of NH_2-UIO-66 (Zr), NH_2-UIO-66 (Zr)@MIP and NH_2-UIO-66 (Zr)@NIP, obtained via the Brunauer–Emmett–Teller (BET) analysis method, were 219 $m^2\ g^{-1}$, 37 $m^2\ g^{-1}$ and 28 $m^2\ g^{-1}$, respectively. Compared to NH_2-UIO-66 (Zr), the surface areas of NH_2-UIO-66 (Zr)@MIP and NH_2-UIO-66 (Zr)@NIP were clearly reduced after coating the imprinted layer [14,35,45]. Meanwhile, there was a slight increase in the surface area of NH_2-UIO-66 (Zr)@MIP in comparison to that of NH_2-UIO-66 (Zr)@NIP, probably due to the existence of imprinted cavities after eliminating the template [46].

The above-mentioned data signified that the FL sensor (NH_2-UIO-66 (Zr)@MIP) was triumphantly fabricated via combining the molecularly imprinted polymer with NH_2-UIO-66 (Zr).

3.3. Optical and Adsorption Properties of NH_2-UIO-66 (Zr)@MIP

The luminescence performance of the developed sensor was firstly studied through FL spectra. As displayed in Figure 2a, NH_2-UIO-66 (Zr)@MIP showed slightly lower FL intensity than NH_2-UIO-66 (Zr) on account of decorating the MIP layer. However, NH_2-UIO-66 (Zr)@MIP exhibited an intense symmetric FL emission peak corresponding to NH_2-UIO-66 (Zr) at 455 nm under 350 nm excitation, which implied that the modification of the surface imprinting layer had no remarkable impact on the FL performance of NH_2-UIO-66 (Zr). As a consequence, the synthesized sensor maintained the unique luminous property of NH_2-UIO-66 (Zr).

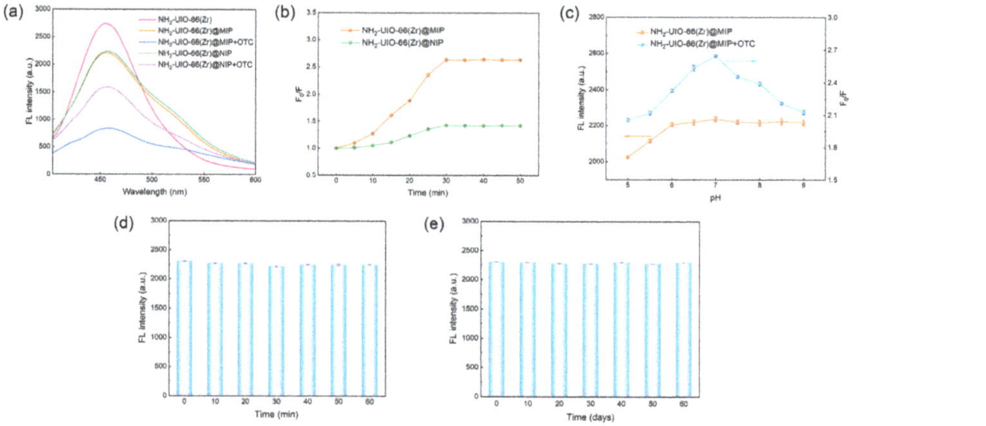

Figure 2. (**a**) FL emission spectra of NH_2-UIO-66 (Zr), NH_2-UIO-66 (Zr)@MIP, NH_2-UIO-66 (Zr)@MIP+OTC, NH_2-UIO-66 (Zr)@NIP and NH_2-UIO-66 (Zr)@NIP+OTC. (**b**) FL response of NH_2-UIO-66 (Zr)@MIP and NH_2-UIO-66 (Zr)@NIP towards OTC at different incubation times. (**c**) Effects of pH values (5–9) on the FL intensity of NH_2-UIO-66 (Zr)@MIP, as well as the FL response between NH_2-UIO-66 (Zr)@MIP and OTC. The influence of (**d**) irradiation time at 350 nm and (**e**) storage period on the FL intensity of NH_2-UIO-66 (Zr)@MIP.

FL experiments measuring the response of the polymers (NH_2-UIO-66 (Zr)@MIP and NH_2-UIO-66 (Zr)@NIP) to OTC were then conducted to explore the practicability of the established sensor (Figure 2a). Clearly, when OTC existed in the detection system, the FL intensity of the polymers at 455 nm was effectively decreased; meanwhile, NH_2-UIO-66 (Zr)@MIP presented a more conspicuous FL change than NH_2-UIO-66 (Zr)@NIP, owing to its better binding capacity and specificity towards OTC. After eliminating OTC, the fluorescence of the polymers was restored and the FL spectrum of NH_2-UIO-66 (Zr)@MIP was basically consistent with that of NH_2-UIO-66 (Zr)@NIP, which indicated that the fabricated NH_2-UIO-66 (Zr)@MIP sensor was viable for the determination of OTC.

Adsorptive kinetics tests of the as-prepared polymers were performed to investigate the adsorption performance of the sensor. As shown in Figure 2b, the FL response (F_0/F) of NH_2-UIO-66 (Zr)@MIP and NH_2-UIO-66 (Zr)@NIP towards the target OTC at different incubation times was recorded. Under the continuously increasing reaction time, the F_0/F value of NH_2-UIO-66 (Zr)@MIP increased quickly at first, reached the maximum at 30 min, and then tended to remain unchanged, which demonstrated that the detection system achieved adsorption equilibrium within 30 min. Therefore, 30 min was considered to be the optimal reacting time in consequent trials. The variation in the trend observed with regard to the FL response of NH_2-UIO-66 (Zr)@NIP was analogous to NH_2-UIO-66 (Zr)@MIP. Nevertheless, NH_2-UIO-66 (Zr)@MIP possessed a higher FL response and adsorption efficiency than NH_2-UIO-66 (Zr)@NIP at the identical reaction time, indicating that the generation of specific identification imprinted sites enhanced the binding affinity of NH_2-UIO-66 (Zr)@MIP towards OTC.

Next, the influences of the pH (5–9) on the fluorescence intensity of NH_2-UIO-66 (Zr)@MIP and the FL response of NH_2-UIO-66 (Zr)@MIP towards OTC were studied in the case of not adding and adding OTC, respectively. As illustrated in Figure 2c, with the pH changing from 5 to 9, the FL intensity of NH_2-UIO-66 (Zr)@MIP first increased, then remained basically stable, and the highest FL intensity was apparently observed at pH = 7. In addition, the maximum FL response value (F_0/F) was concurrently obtained at pH = 7, proving that NH_2-UIO-66 (Zr)@MIP possessed the strongest adsorption binding efficiency to OTC. Thus, pH = 7 was chosen as the best pH for the sensing system in subsequent FL detection experiments.

Finally, the fluorescent stability of the obtained sensor was estimated by measuring the FL intensity at 455 nm of NH_2-UIO-66 (Zr)@MIP under diverse ultraviolet (UV) light radiation times and storing times. As shown in Figure 2d, NH_2-UIO-66 (Zr)@MIP exhibited an unapparent FL change after UV irradiation at 350 nm for 60 min, signifying the exceptional photo-bleaching resistance of the proposed NH_2-UIO-66 (Zr)@MIP-based sensor. At the same time, the FL intensity of NH_2-UIO-66 (Zr)@MIP remained steady within 60 days (Figure 2e), which indicated the superior storage stability of the resulting sensor.

The above experimental results confirm that the NH_2-UIO-66 (Zr)@MIP-based sensor had good optical properties and a good adsorption performance.

3.4. Construction of FL Detection System for OTC

Under the above optimum sensing conditions, the concentration responses of the as-synthesized FL polymers for OTC were assessed by collecting and analyzing the FL spectra of NH_2-UIO-66 (Zr)@MIP and NH_2-UIO-66 (Zr)@NIP after the addition of different levels of OTC. As portrayed in Figure 3a, by increasing the OTC concentration from 0.05 to 40 μg mL^{-1}, the FL intensity of NH_2-UIO-66 (Zr)@MIP was progressively quenched. Clearly, NH_2-UIO-66 (Zr)@NIP showed an analogous FL change trend towards NH_2-UIO-66 (Zr)@MIP. Nevertheless, compared with NH_2-UIO-66 (Zr)@NIP, NH_2-UIO-66 (Zr)@MIP presented a more noticeable FL quenching response when the same concentration of OTC was added (Figure 3b), which was the result of the presence of molecularly imprinted sites matching OTC in NH_2-UIO-66 (Zr)@MIP. Furthermore, the relationship between the FL quenching signal (F_0/F) and the concentration of OTC can be described via the Stern–Volmer equation [14,47]:

$$F_0/F = 1 + K_{SV}C_{OTC}$$

where C_{OTC} and K_{SV} are the concentration of OTC and the FL quenching constant, respectively. F_0 and F represent the FL intensity of the sensing system before and after adding OTC. The imprinting factor (*IF*), set as the specific value of $K_{SV,\ NH2\text{-}UIO\text{-}66\ (Zr)@MIP}$ to $K_{SV,\ NH2\text{-}UIO\text{-}66\ (Zr)@NIP}$, was used as a key index to estimate the adsorption binding capacity of the NH_2-UIO-66 (Zr)@MIP sensor towards OTC.

Figure 3c shows that there was a linear dependence between the FL quenching signal of NH_2-UIO-66 (Zr)@MIP and the OTC concentration (0.05–40 μg mL^{-1}). The limit of detection (LOD) of 0.012 μg mL^{-1} was achieved on the basis of the standard $3S_b/slope$ ($S_b = 1.63 \times 10^{-4}$, was the standard deviation of blank measurements, $n = 9$) [22,48]. Plainly, NH_2-UIO-66 (Zr)@MIP had a more notable FL response change than NH_2-UIO-66 (Zr)@NIP at an identical OTC dose, which emphasized the outstanding selectivity and adsorptive efficiency of NH_2-UIO-66 (Zr)@MIP towards OTC. Simultaneously, the good specific recognition capacity of NH_2-UIO-66 (Zr)@MIP was also authenticated by comparing the slopes of the corresponding standard linear equations of NH_2-UIO-66 (Zr)@MIP ($F_0/F = 0.0399C_{OTC} + 1.0296$, $R^2 = 0.998$) and NH_2-UIO-66 (Zr)@NIP ($F_0/F = 0.0098C_{OTC} + 1.0275$, $R^2 = 0.994$) (Figure 3c,d). Moreover, the maximum *IF* value was calculated as 3.95, further revealing that the NH_2-UIO-66 (Zr)@MIP sensor had a superior selective identification capacity and superior adsorption affinity for OTC.

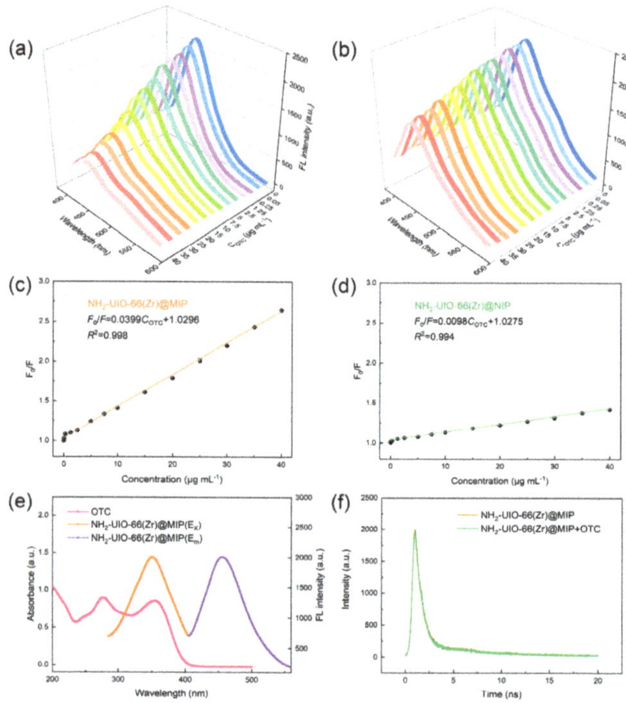

Figure 3. FL spectra of (**a**) NH_2-UIO-66 (Zr)@MIP and (**b**) NH_2-UIO-66 (Zr)@NIP under different concentrations of OTC (excited at 350 nm). The Stern–Volmer plots for NH_2-UIO-66 (Zr)@MIP (**c**) and NH_2-UIO-66 (Zr)@NIP (**d**). (**e**) The UV absorption spectrum of OTC and the FL spectra of NH_2-UIO-66 (Zr)@MIP (Ex, FL excitation spectrum; Em, FL emission spectrum). (**f**) FL emission decay curves of NH_2-UIO-66 (Zr)@MIP alone and NH_2-UIO-66 (Zr)@MIP with 40 μg mL^{-1} concentrations of OTC.

The potential FL sensing mechanism of the NH_2-UIO-66 (Zr)@MIP sensor towards the OTC molecule was investigated and discussed. The FL spectrum of NH_2-UIO-66 (Zr)@MIP and the UV absorption spectrum of OTC are illustrated in Figure 3e. Evidently, the FL emission spectrum of NH_2-UIO-66 (Zr)@MIP almost did not overlap with the UV absorption spectrum of OTC, but there was a partial and effective overlap between the excitation spectrum of NH_2-UIO-66 (Zr)@MIP and the UV absorption band of OTC. Therefore, the possible mechanism was initially determined as the inner filter effect (IFE), rather than FL resonance energy transfer (FRET) [37,49]. To be specific, the FL lifetime of the fluorescent material decreases in the FRET process, while it remains invariant if IFE occurs [50,51]. Accordingly, to further elucidate the FL quenching phenomenon, the FL lifetime of NH_2-UIO-66 (Zr)@MIP before and after OTC adsorption was analyzed. The FL emission decay curves are presented in Figure 3f. It is noteworthy that the FL lifetime of NH_2-UIO-66 (Zr)@MIP showed a slight variation from 3.61 ns to 3.48 ns after OTC treatment, further demonstrating that the OTC-initiated FL quenching behavior was mainly caused by IFE.

3.5. Selectivity and Specificity Analysis

The selectivity of the constructed FL sensing system was assessed by recording and contrasting the FL changes in the polymers to OTC and its structural analogues (TET, DOX, CTC and CAP). The chemical structures of these mentioned antibiotics, which are first portrayed in Figure 4a, enable the specific recognition behavior of the NH_2-UIO-66 (Zr)@MIP-based sensor towards the OTC molecule to be better understood. As presented in Figure 4b, the influence of OTC on the FL quenching degree of NH_2-UIO-66 (Zr)@MIP was notably greater than that of other antibiotics with a similar structure, which was mainly owing to the formation of the special recognition cavities in NH_2-UIO-66 (Zr)@MIP that utterly conformed the size, shape and functional group of OTC during the molecular imprinting process. Because there were no specific recognition sites corresponding to TET, DOX, CTC and CAP in NH_2-UIO-66 (Zr)@MIP, the NH_2-UIO-66 (Zr)@MIP sensor had a comparatively small FL response value (F_0/F) to these analogs (Figure 4b). Therefore, it was difficult for NH_2-UIO-66 (Zr)@MIP to selectively capture and effectively bind with these analogues in the detection process. Furthermore, compared with NH_2-UIO-66 (Zr)@MIP, NH_2-UIO-66 (Zr)@NIP, without specific recognition sites, generated a relatively low FL response to OTC and other analogues, which was attributed to the nonspecific adsorption between NH_2-UIO-66 (Zr)@NIP and these analytes. The above results suggest that the introduction of highly selective imprinted sites endowed the NH_2-UIO-66 (Zr)@MIP sensor with the ability to specifically identify OTC.

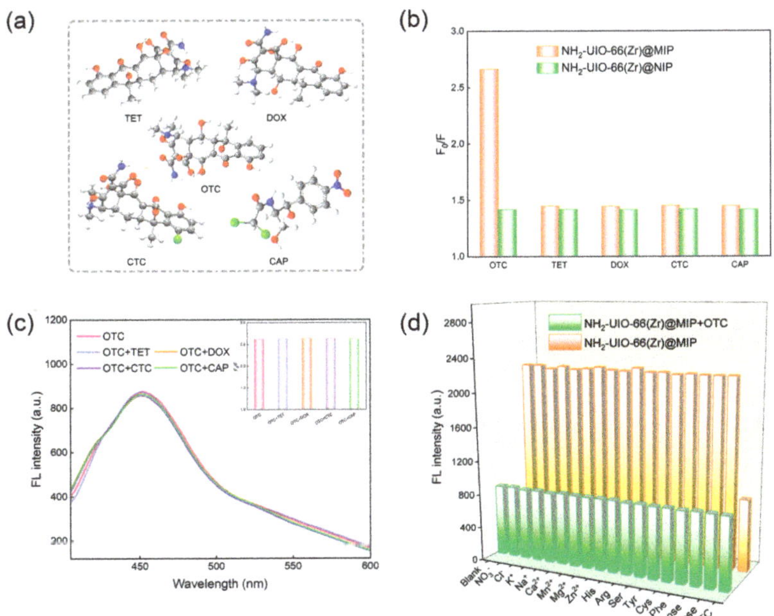

Figure 4. (**a**) The chemical structural formulas of OTC, TET, DOX, CTC and CAP. (**b**) The FL response (F_0/F) of NH_2-UIO-66 (Zr)@MIP and NH_2-UIO-66 (Zr)@NIP to 40 μg mL^{-1} of OTC, TET, DOX, CTC and CAP. (**c**) The influence of TET, DOX, CTC and CAP (80 μg mL^{-1}) on the combination of OTC (40 μg mL^{-1}) and NH_2-UIO-66 (Zr)@MIP. (**d**) The influence of coexisting substances (200 μg mL^{-1}) on the FL change of NH_2-UIO-66 (Zr)@MIP in the absence or presence of OTC (40 μg mL^{-1}).

The FL response changes (F_0/F) of NH_2-UIO-66 (Zr)@MIP to different concentrations of structural analogues were also studied to investigate the selectivity of the established sensor. As shown in Figure S5, in the absence of OTC, the FL detection system exhibited slight differences in FL changes for different analogues, but the FL changes (F_0/F) apparently increased after introducing OTC. In addition, there was no significant linear relationship between the concentration of the analogue and FL change (F_0/F), which indicated that the constructed FL detection system had good selectivity for the target OTC.

In order to further assess the ability of NH_2-UIO-66 (Zr)@MIP to specifically recognize OTC, a competitive adsorption test was conducted by adding the possible coexisting structural analogs (TET, DOX, CTC and CAP) into the mixed system containing NH_2-UIO-66 (Zr)@MIP and OTC. As revealed in Figure 4c, NH_2-UIO-66 (Zr)@MIP displayed tiny FL variation even though the concentration of each analogue was twice that of OTC, demonstrating that the interference caused by the presence of the coexisting analogues in the sensing system was inconspicuous. The good specificity of NH_2-UIO-66 (Zr)@MIP towards OTC was probably due to the existence of OTC-imprinted sites generated via hydrogen bond interaction, which improved the combination efficiency of the sensor to OTC, and inhibited the nonspecific adsorption of the coexisting analogs. Thus, the obtained sensor possessed gratifying selectivity and specificity for the target OTC.

Due to the matrix complexity of the practical samples, the potential coexisting interfering substances in the real detection process, such as sugars (glucose and sucrose), ions (Cl^-, NO_3^-, K^+, Na^+, Ca^{2+}, Zn^{2+}, Mg^{2+}, and Mn^{2+}) and common amino acids (Arg, Cys, Phe, Tyr, His and Ser), will directly influence the accuracy of the test results. Therefore, the ability of the NH_2-UIO-66 (Zr)@MIP sensor to avoid interference was explored by introducing the above interfering analytes into the constructed FL sensing system in the presence and absence of OTC. As shown in Figure 4d, the concentration of the added interfering species was up to 200 μg mL^{-1} (5 times of OTC concentration), but the influence of these interfering substances on the FL intensity of NH_2-UIO-66 (Zr)@MIP was minor, which was in accordance with the blank experimental group. After adding OTC, the FL of the sensing system was significantly quenched, indicating that these interfering species had a negligible effect on the FL detection of OTC. The above results indicated that the developed FL sensor exhibited excellent selectivity and specificity towards OTC, as well as superior anti-interference characteristics in a complicated environment, thus providing a feasible tool for the FL detection of OTC.

3.6. *Real Sample Analysis*

The commercial milk bought from the local supermarket was taken as a practical sample in order to analyze and authenticate the applicability of the FL sensing approach created in this study. No OTC was found in the blank milk samples when using the NH_2-UIO-66 (Zr)@MIP method and HPLC means. Consequently, the spiking and recovery tests were performed by introducing different levels of OTC standards (0, 1, 15 and 25 μg mL^{-1}) into the milk samples, and the relevant analysis results are presented in Table 1. The recovery rates of the milk samples obtained by using the established optical sensing system were between 93.56% and 98.21%, and the relative standard deviation (RSD) was <2.59%, which was comparable to the results of the HPLC method. These satisfactory data confirmed that the fabricated sensor was practicable for quantitatively detecting OTC in milk samples.

3.7. *Accuracy Assessment of the Established Fluorescence Method*

The intra-day and inter-day precisions of the established NH_2-UIO-66 (Zr)@MIP sensing method were also appraised by detecting the spiked samples using different concentrations of OTC standards (1, 15 and 25 μg mL^{-1}). As shown in Table S4, the corresponding RSD values were lower than 3.05% and 4.07% (n = 6), respectively, indicating that the sensor had sterling reproducibility as well as good accuracy, and was suitable for the determination of OTC in real products.

Table 1. Detection of OTC in milk samples using the NH_2-UIO-66 (Zr)@MIP sensor and HPLC.

Sample	Spiked ($\mu g\ mL^{-1}$)	NH_2-UIO-66 (Zr)@MIP			HPLC		
		Found ($\mu g\ mL^{-1}$)	Recovery (%)	RSD (%, n = 3)	Found ($\mu g\ mL^{-1}$)	Recovery (%)	RSD (%, n = 3)
Milk 1	0.00	0.00	-	-	0.00	-	-
	1.00	0.95 ± 0.02	95.38 ± 2.47	2.59	0.96 ± 0.02	96.16 ± 2.50	2.60
	15.00	14.50 ± 0.29	96.65 ± 1.96	2.03	14.73 ± 0.10	98.23 ± 0.66	0.67
	25.00	24.45 ± 0.16	97.79 ± 0.65	0.67	24.54 ± 0.13	98.14 ± 0.51	0.52
Milk 2	0.00	0.00	-	-	0.00	-	-
	1.00	0.94 ± 0.02	93.56 ± 1.86	1.98	0.97 ± 0.02	96.68 ± 1.88	1.94
	15.00	14.67 ± 0.14	97.78 ± 0.91	0.93	14.78 ± 0.04	98.51 ± 0.26	0.27
	25.00	24.33 ± 0.11	97.32 ± 0.43	0.44	24.52 ± 0.12	98.10 ± 0.47	0.48
Milk 3	0.00	0.00	-	-	0.00	-	-
	1.00	0.94 ± 0.01	93.67 ± 1.00	1.07	0.96 ± 0.01	95.59 ± 1.10	1.15
	15.00	14.73 ± 0.17	98.21 ± 1.10	1.12	14.64 ± 0.01	97.62 ± 0.04	0.04
	25.00	24.41 ± 0.16	97.65 ± 0.64	0.65	24.61 ± 0.09	98.45 ± 0.35	0.35

3.8. Approach Performance Comparison

The performance of the designed sensing assay was compared to some subsistent FL-based methods for OTC detection, the results of which are shown in Table S5. Visibly, the fluorescent sensing approach established in this study exhibited a relatively wide linear range (0.05–40 $\mu g\ mL^{-1}$), a lower LOD (0.012 $\mu g\ mL^{-1}$), a comparable recovery rate (93.56–98.21%) and an acceptable accuracy (RSD < 2.59%). The good performance of the NH_2-UIO-66 (Zr)@MIP-based sensor came from the following aspects. On one hand, introducing NH_2-UIO-66 (Zr), which has sterling FL characteristics and a large surface area, as the support improved the sensitivity of the proposed FL sensor. On the other hand, the integration of the surface-imprinted layer endowed the sensor with the ability to specifically identify and accurately absorb the target from the complicated matrix, which helped to avoid the tedious sample pretreatment process and to reduce the analysis time. Taken together, the constructed NH_2-UIO-66 (Zr)@MIP-based sensor can be appropriately applied in order to quantitatively monitor trace OTC in a sophisticated system.

4. Conclusions

In summary, a novel turn-off fluorescent biomimetic sensor (NH_2-UIO-66 (Zr)@MIP) was successfully designed based on an amino-functionalized zirconium (IV) metal–organic framework embedded in a molecularly imprinted polymer in order to selectively recognize and sensitively detect trace OTC. Because of the combination of MIP, with an excellent specific recognition ability, and NH_2-UIO-66 (Zr), with unique fluorescent characteristics, the fabricated NH_2-UIO-66 (Zr)MIP sensor showed exceptional sensitivity, excellent selectivity, desirable anti-interference capacity, as well as good precision during the process of detecting OTC. Additionally, the constructed FL sensing system was triumphantly employed in order to determine OTC in milk with satisfying results, displaying good practical application prospects. Therefore, it is expected that the proposed NH_2-UIO-66 (Zr)@MIP-based sensor will provide a significant detection platform for OTC residue analysis in food safety evaluation.

Supplementary Materials: The following supporting information can be downloaded at: https://www.mdpi.com/article/10.3390/foods12112255/s1, Refs. [52–56], Instruments, The detection conditions and sample analysis procedures for HPLC method, Optimization of synthesis conditions for NH_2-UIO-66 (Zr)@MIP, Figure S1: Diameter distribution of NH_2-UIO-66 (Zr); Figure S2: EDS image of NH_2-UIO-66 (Zr); Figure S3: Diameter distribution of NH_2-UIO-66 (Zr)@MIP (a) and NH_2-UIO-66 (Zr)@NIP (b); Figure S4: The high-angle annular dark-field (HAADF) image of NH_2-UIO-66 (Zr)@MIP, and the corresponding element mappings of C, N, O and Zr; Figure S5: The FL response (F_0/F) of NH_2-UIO-66 (Zr)@MIP to different concentrations (2.5–40 $\mu g\ mL^{-1}$) of OTC, TET,

DOX, CTC and CAP, as well as to 40 μg mL^{-1} OTC + 40 μg mL^{-1} TET/DOX/CTC/CAP; Table S1: Optimization of the molar ratio of OTC, MAA and EGDMA; Table S2: Optimization of NH_2-UIO-66 (Zr) addition; Table S3: Optimization of polymerization time; Table S4: The intra- and inter-day precisions of the designed FL assay (n = 6); Table S5: Comparison of the constructed NH_2-UIO-66 (Zr)@MIP-based sensing approach with other reported FL methods for OTC detection in milk.

Author Contributions: Conceptualization, X.W.; methodology, X.W.; software, X.W.; validation, X.W.; formal analysis, X.W. and C.L.; investigation, Y.C. and L.C.; data curation, X.W. and H.W.; writing—original draft preparation, X.W.; writing—review and editing, G.F.; supervision, G.F. All authors have read and agreed to the published version of the manuscript.

Funding: This research was funded by The Project Program of Key Laboratory of Tianjin Key Laboratory of Food Quality and Health, China (project No. TJS202201).

Data Availability Statement: The data presented in this study are available on request from the corresponding author.

Conflicts of Interest: The authors declare no conflict of interest.

References

1. Lustig, W.P.; Mukherjee, S.; Rudd, N.D.; Desai, A.V.; Li, J.; Ghosh, S.K. Metal-organic frameworks: Functional luminescent and photonic materials for sensing applications. *Chem. Soc. Rev.* **2017**, *46*, 3242–3285. [CrossRef] [PubMed]
2. Cui, Y.J.; Yue, Y.F.; Qian, G.D.; Chen, B.L. Luminescent Functional Metal-Organic Frameworks. *Chem. Rev.* **2012**, *112*, 1126–1162. [CrossRef] [PubMed]
3. Bai, Y.; Dou, Y.B.; Xie, L.H.; Rutledge, W.; Li, J.R.; Zhou, H.C. Zr-based metal-organic frameworks: Design, synthesis, structure, and applications. *Chem. Soc. Rev.* **2016**, *45*, 2327–2367. [CrossRef] [PubMed]
4. Kandiah, M.; Nilsen, M.H.; Usseglio, S.; Jakobsen, S.; Olsbye, U.; Tilset, M.; Larabi, C.; Quadrelli, E.A.; Bonino, F.; Lillerud, K.P. Synthesis and Stability of Tagged UiO-66 Zr-MOFs. *Chem. Mater.* **2010**, *22*, 6632–6640. [CrossRef]
5. Wang, N.; Xie, M.G.; Wang, M.K.; Li, Z.X.; Su, X.G. UiO-66-NH_2 MOF-based ratiometric fluorescent probe for the detection of dopamine and reduced glutathione. *Talanta* **2020**, *220*, 7. [CrossRef]
6. Zhu, H.L.; Huang, J.P.; Zhou, Q.Y.; Lv, Z.W.; Li, C.Y.; Hu, G. Enhanced luminescence of NH_2-UiO-66 for selectively sensing fluoride anion in water medium. *J. Lumin.* **2019**, *208*, 67–74. [CrossRef]
7. Yang, J.; Dai, Y.; Zhu, X.Y.; Wang, Z.; Li, Y.S.; Zhuang, Q.X.; Shi, J.L.; Gu, J.L. Metal-organic frameworks with inherent recognition sites for selective phosphate sensing through their coordination-induced fluorescence enhancement effect. *J. Mater. Chem. A* **2015**, *3*, 7445–7452. [CrossRef]
8. Wang, X.H.; Wang, X.F. UiO-66-NH_2 based fluorescent sensing for detection of tetracyclines in milk. *RSC Adv.* **2022**, *12*, 23427–23436. [CrossRef]
9. Zhang, H.F.; Kang, Z.W.; Zhu, H.; Lin, H.T.; Yang, D.P. ZnO/C nanocomposite grafted molecularly imprinted polymers as photoelectrochemical sensing interface for ultrasensitive and selective detection of chloramphenicol. *Sci. Total Environ.* **2023**, *859*, 160284. [CrossRef]
10. Hatamluyi, B.; Hashemzadeh, A.; Darroudi, M. A novel molecularly imprinted polymer decorated by CQDs@HBNNS nanocomposite and UiO-66-NH_2 for ultra-selective electrochemical sensing of Oxaliplatin in biological samples. *Sens. Actuators B-Chem.* **2020**, *307*, 127614. [CrossRef]
11. Refaat, D.; Aggour, M.G.; Farghali, A.A.; Mahajan, R.; Wiklander, J.G.; Nicholls, I.A.; Piletsky, S.A. Strategies for Molecular Imprinting and the Evolution of MIP Nanoparticles as Plastic Antibodies-Synthesis and Applications. *Int. J. Mol. Sci.* **2019**, *20*, 6304. [CrossRef]
12. Cowen, T.; Cheffena, M. Template Imprinting Versus Porogen Imprinting of Small Molecules: A Review of Molecularly Imprinted Polymers in Gas Sensing. *Int. J. Mol. Sci.* **2022**, *23*, 9642. [CrossRef]
13. Afshar, E.A.; Taher, M.A.; Karimi, F.; Karaman, C.; Moradi, O. Ultrasensitive and highly selective "turn-on" fluorescent sensor for the detection and measurement of melatonin in juice samples. *Chemosphere* **2022**, *295*, 133869. [CrossRef] [PubMed]
14. Amiripour, F.; Ghasemi, S.; Azizi, S.N. Fo euro rster resonance energy transfer-based molecularly imprinted polymer/amine-functionalized metal-organic framework nanocomposite for trace level detection of 4-nitrophenol. *Anal. Chim. Acta* **2022**, *1202*, 339638. [CrossRef]
15. Amiripour, F.; Ghasemi, S.; Azizi, S.N. Design of turn-on luminescent sensor based on nanostructured molecularly imprinted polymer-coated zirconium metal-organic framework for selective detection of chloramphenicol residues in milk and honey. *Food Chem.* **2021**, *347*, 129034. [CrossRef] [PubMed]
16. Gaudin, V. Advances in biosensor development for the screening of antibiotic residues in food products of animal origin—A comprehensive review. *Biosens. Bioelectron.* **2017**, *90*, 363–377. [CrossRef] [PubMed]
17. Marshall, B.M.; Levy, S.B. Food Animals and Antimicrobials: Impacts on Human Health. *Clin. Microbiol. Rev.* **2011**, *24*, 718–733. [CrossRef]

18. Chopra, I.; Roberts, M. Tetracycline antibiotics: Mode of action, applications, molecular biology, and epidemiology of bacterial resistance. *Microbiol. Mol. Biol. Rev.* **2001**, *65*, 232. [CrossRef]
19. Singer, R.S.; Finch, R.; Wegener, H.C.; Bywater, R.; Walters, J.; Lipsitch, M. Antibiotic resistance—The interplay between antibiotic use in animals and human beings. *Lancet Infect. Dis.* **2003**, *3*, 47–51. [CrossRef]
20. Wang, J.X.; Zou, L.H.; Xu, J.J.; Zhang, R.; Zhang, H.B. Molecularly imprinted fluoroprobes doped with Ag nanoparticles for highly selective detection of oxytetracycline in real samples. *Anal. Chim. Acta* **2021**, *1161*, 338326. [CrossRef]
21. Hirsch, R.; Ternes, T.; Haberer, K.; Kratz, K.L. Occurrence of antibiotics in the aquatic environment. *Sci. Total Environ.* **1999**, *225*, 109–118. [CrossRef]
22. Li, C.H.; Zhu, L.; Yang, W.X.; He, X.; Zhao, S.L.; Zhang, X.S.; Tang, W.Z.; Wang, J.L.; Yue, T.L.; Li, Z.H. Amino-Functionalized Al-MOF for Fluorescent Detection of Tetracyclines in Milk. *J. Agric. Food Chem.* **2019**, *67*, 1277–1283. [CrossRef]
23. Bai, X.Y.; Zhang, Y.; Gao, W.K.; Zhao, D.Y.; Yang, D.P.; Jia, N.Q. Hollow ZnS-CdS nanocage based photoelectrochemical sensor combined with molecularly imprinting technology for sensitive detection of oxytetracycline. *Biosens. Bioelectron.* **2020**, *168*, 112522. [CrossRef]
24. Zhao, M.; Wang, J.T.; Lian, Z.R. Fluorescence assay of oxytetracycline in seawater after selective capture using magnetic molecularly imprinted nanoparticles. *Mar. Pollut. Bull.* **2021**, *163*, 111962. [CrossRef]
25. Mu, G.F.; Liu, H.T.; Xu, L.N.; Tian, L.H.; Luan, F. Matrix Solid-Phase Dispersion Extraction and Capillary Electrophoresis Determination of Tetracycline Residues in Milk. *Food Anal. Meth.* **2012**, *5*, 148–153. [CrossRef]
26. Gajda, A.; Posyniak, A. Liquid chromatography—Tandem mass spectrometry method for the determination of ten tetracycline residues in muscle samples. *Bull. Vet. Inst. Pulawy* **2015**, *59*, 345–352. [CrossRef]
27. Chen, Y.N.; Kong, D.Z.; Liu, L.Q.; Song, S.S.; Kuang, H.; Xu, C.L. Development of an ELISA and Immunochromatographic Assay for Tetracycline, Oxytetracycline, and Chlortetracycline Residues in Milk and Honey Based on the Class-Specific Monoclonal Antibody. *Food Anal. Meth.* **2016**, *9*, 905–914. [CrossRef]
28. Liu, X.; Zhang, Z.L.; Peng, J.D.; He, Y.T. High-performance liquid chromatography with resonance Rayleigh scattering detection for determining four tetracycline antibiotics. *Anal. Methods* **2014**, *6*, 9361–9366. [CrossRef]
29. Fu, Y.Z.; Huang, L.; Zhao, S.J.; Xing, X.J.; Lan, M.H.; Song, X.Z. A carbon dot-based fluorometric probe for oxytetracycline detection utilizing a Forster resonance energy transfer mechanism. *Spectrochim. Acta Part A-Mol. Biomol. Spectrosc.* **2021**, *246*, 6. [CrossRef] [PubMed]
30. Xing, X.J.; Huang, L.; Zhao, S.J.; Xiao, J.F.; Lan, M.H. S,N-Doped carbon dots for tetracyclines sensing with a fluorometric spectral response. *Microchem. J.* **2020**, *157*, 105065. [CrossRef]
31. Satana Kara, H.E.; Demirhan, B.; Er Demirhan, B. Highly luminescent water-dispersed silicon quantum dots for fluorometric determination of oxytetracycline in milk samples. *Turk. J. Chem.* **2020**, *44*, 1713–1722. [CrossRef] [PubMed]
32. Wang, W.J.; Xu, Y.Q.; Liu, X.Q.; Peng, L.; Huang, T.; Yan, Y.S.; Li, C.X. Efficient fabrication of ratiometric fluorescence imprinting sensors based on organic-inorganic composite materials and highly sensitive detection of oxytetracycline in milk. *Microchem. J.* **2020**, *157*, 105053. [CrossRef]
33. Hu, X.L.; Guo, Y.; Wang, T.; Liu, C.; Yang, Y.K.; Fang, G.Z. A selectivity-enhanced ratiometric fluorescence imprinted sensor based on synergistic effect of covalent and non-covalent recognition units for ultrasensitive detection of ribavirin. *J. Hazard. Mater.* **2022**, *421*, 126748. [CrossRef] [PubMed]
34. Eskandari, H.; Amirzehni, M.; Asadollahzadeh, H.; Hassanzadeh, J.; Eslami, P.A. MIP-capped terbium MOF-76 for the selective fluorometric detection of cefixime after its preconcentration with magnetic graphene oxide. *Sens. Actuators B-Chem.* **2018**, *275*, 145–154. [CrossRef]
35. Hu, X.L.; Guo, Y.; Zhang, J.N.; Wang, X.H.; Fang, G.Z.; Wang, S. A signal-amplified ratiometric fluorescence biomimetic sensor based on the synergistic effect of IFE and AE for the visual smart monitoring of oxytetracycline. *Chem. Eng. J.* **2022**, *433*, 134499. [CrossRef]
36. Qian, S.H.; Qiao, L.N.; Xu, W.X.; Jiang, K.; Wang, Y.H.; Lin, H.W. An inner filter effect-based near-infrared probe for the ultrasensitive detection of tetracyclines and quinolones. *Talanta* **2019**, *194*, 598–603. [CrossRef]
37. Hu, X.L.; Cao, Y.C.; Cai, L.; Wang, H.Y.; Fang, G.Z.; Wang, S. A smartphone-assisted optosensing platform based on chromium-based metal-organic framework signal amplification for ultrasensitive and real-time determination of oxytetracycline. *J. Hazard. Mater.* **2023**, *444*, 130395. [CrossRef]
38. Chen, Q.; He, Q.Q.; Lv, M.M.; Xu, Y.L.; Yang, H.B.; Liu, X.T.; Wei, F.Y. Selective adsorption of cationic dyes by UiO-66-NH_2. *Appl. Surf. Sci.* **2015**, *327*, 77–85. [CrossRef]
39. Fang, X.; Wu, S.B.; Wu, Y.H.; Yang, W.; Li, Y.L.; He, J.Y.; Hong, P.D.; Nie, M.X.; Xie, C.; Wu, Z.J.; et al. High-efficiency adsorption of norfloxacin using octahedral UIO-66-NH_2 nanomaterials: Dynamics, thermodynamics, and mechanisms. *Appl. Surf. Sci.* **2020**, *518*, 146226. [CrossRef]
40. Lin, K.Y.A.; Liu, Y.T.; Chen, S.Y. Adsorption of fluoride to UiO-66-NH_2 in water: Stability, kinetic, isotherm and thermodynamic studies. *J. Colloid Interface Sci.* **2016**, *461*, 79–87. [CrossRef]
41. Li, C.M.; Huang, J.P.; Zhu, H.L.; Liu, L.L.; Feng, Y.M.; Hu, G.; Yu, X.B. Dual-emitting fluorescence of Eu/Zr-MOF for ratiometric sensing formaldehyde. *Sens. Actuators B-Chem.* **2017**, *253*, 275–282. [CrossRef]

42. Hu, X.L.; Zhao, Y.Q.; Dong, J.Y.; Liu, C.; Qi, Y.; Fang, G.Z.; Wang, S. A strong blue fluorescent nanoprobe based on Mg/N co-doped carbon dots coupled with molecularly imprinted polymer for ultrasensitive and highly selective detection of tetracycline in animal-derived foods. *Sens. Actuators B-Chem.* **2021**, *338*, 129809. [CrossRef]
43. Han, S.; Yao, A.X.; Ding, Y.X.; Leng, Q.X.; Teng, F.; Zhao, L.; Sun, R.N.; Bu, H.Z. A dual-template imprinted polymer based on amino-functionalized zirconium-based metal-organic framework for delivery of doxorubicin and phycocyanin with synergistic anticancer effect. *Eur. Polym. J.* **2022**, *170*, 111161. [CrossRef]
44. Liang, Y.T.; He, J.; Huang, Z.P.; Li, H.Y.; Zhang, Y.X.; Wang, H.G.; Rui, C.F.; Li, Y.Y.; You, L.Q.; Li, K.; et al. An amino-functionalized zirconium-based metal-organic framework of type UiO-66-NH_2 covered with a molecularly imprinted polymer as a sorbent for the extraction of aflatoxins AFB1, AFB2, AFG1 and AFG2 from grain. *Microchim. Acta* **2020**, *187*, 32. [CrossRef] [PubMed]
45. Yang, J.Y.; Feng, W.B.; Liang, K.S.; Chen, C.Y.; Cai, C.Q. A novel fluorescence molecularly imprinted sensor for Japanese encephalitis virus detection based on metal organic frameworks and passivation-enhanced selectivity. *Talanta* **2020**, *212*, 120744. [CrossRef] [PubMed]
46. Xu, L.H.; Pan, M.F.; Fang, G.Z.; Wang, S. Carbon dots embedded metal-organic framework@molecularly imprinted nanoparticles for highly sensitive and selective detection of quercetin. *Sens. Actuators B-Chem.* **2019**, *286*, 321–327. [CrossRef]
47. Hu, X.L.; Cao, Y.C.; Tian, Y.Y.; Qi, Y.; Fang, G.Z.; Wang, S. A molecularly imprinted fluorescence nanosensor based on upconversion metal-organic frameworks for alpha-cypermethrin specific recognition. *Microchim. Acta* **2020**, *187*, 632. [CrossRef]
48. Zeng, L.S.; Zhang, X.; Wang, X.; Cheng, D.Q.; Li, R.F.; Han, B.; Wu, M.M.; Zhuang, Z.J.; Ren, A.N.; Zhou, Y.K.; et al. Simultaneous fluorescence determination of bisphenol A and its halogenated analogs based on a molecularly imprinted paper-based analytical device and a segment detection strategy. *Biosens. Bioelectron.* **2021**, *180*, 113106. [CrossRef]
49. Zhou, Y.; Yang, Q.; Zhang, D.N.; Gan, N.; Li, Q.P.; Cuan, J. Detection and removal of antibiotic tetracycline in water with a highly stable luminescent MOF. *Sens. Actuators B-Chem.* **2018**, *262*, 137–143. [CrossRef]
50. Zhang, Y.Q.; Wu, X.H.; Mao, S.; Tao, W.Q.; Li, Z. Highly luminescent sensing for nitrofurans and tetracyclines in water based on zeolitic imidazolate framework-8 incorporated with dyes. *Talanta* **2019**, *204*, 344–352. [CrossRef]
51. Tan, H.L.; Wu, X.Y.; Weng, Y.H.; Lu, Y.J.; Huang, Z.Z. Self-Assembled FRET Nanoprobe with Metal-Organic Framework As a Scaffold for Ratiometric Detection of Hypochlorous Acid. *Anal. Chem.* **2020**, *92*, 3447–3454. [CrossRef] [PubMed]
52. Xu, N.; Yuan, Y.Q.; Yin, J.H.; Wang, X.; Meng, L. One-pot hydrothermal synthesis of luminescent silicon-based nanoparticles for highly specific detection of oxytetracycline via ratiometric fluorescent strategy. *RSC Adv.* **2017**, *7*, 48429–48436. [CrossRef]
53. Xu, Y.Q.; Huang, T.; Wang, S.O.; Yan, Y.S. Mesoporous silica-based molecularly imprinted fluorescence sensor for the ultrafast and sensitive recognition of oxytetracycline. *J. Food Compos. Anal.* **2022**, *108*, 9. [CrossRef]
54. Cai, Z.F.; Li, H.Y.; Wang, X.S.; Min, C.; Wen, J.Q.; Fu, R.X.; Dai, Z.Y.; Chen, J.; Guo, M.Z.; Yang, H.J.; et al. Highly luminescent copper nanoclusters as temperature sensors and "turn off" detection of oxytetracycline. *Colloid Surf. A-Physicochem. Eng. Asp.* **2022**, *647*, 10. [CrossRef]
55. Wu, H.F.; Xu, M.Q.; Chen, Y.B.; Zhang, H.L.; Shen, Y.J.; Tang, Y.F. A Highly Sensitive and Selective Nano-Fluorescent Probe for Ratiometric and Visual Detection of Oxytetracycline Benefiting from Dual Roles of Nitrogen-Doped Carbon Dots. *Nanomaterials* **2022**, *12*, 4306. [CrossRef]
56. Li, Y.; Wang, Y.; Du, P.Y.; Zhang, L.B.; Liu, Y.; Lu, X.Q. Fabrication of carbon dots@hierarchical mesoporous ZIF-8 for simultaneous ratiometric fluorescence detection and removal of tetracycline antibiotics. *Sens. Actuator B-Chem.* **2022**, *358*, 8. [CrossRef]

MDPI
St. Alban-Anlage 66
4052 Basel
Switzerland
www.mdpi.com

Foods Editorial Office
E-mail: foods@mdpi.com
www.mdpi.com/journal/foods

www.ingramcontent.com/pod-product-compliance
Lightning Source LLC
Chambersburg PA
CBHW041951260926
51850CB00019B/2620
9783725805952